SUBREGIONALISM AND WORLD ORDER

Subregionalism and World Order

Edited by

Glenn Hook
Professor of Japanese Studies
School of East Asian Studies
University of Sheffield

and

Ian Kearns
Lecturer in Politics
Department of Politics
University of Sheffield

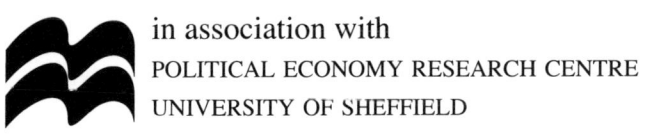 in association with
POLITICAL ECONOMY RESEARCH CENTRE
UNIVERSITY OF SHEFFIELD

First published in Great Britain 1999 by
MACMILLAN PRESS LTD
Houndmills, Basingstoke, Hampshire RG21 6XS and London
Companies and representatives throughout the world

A catalogue record for this book is available from the British Library.

ISBN 0-333-71960-3

 First published in the United States of America 1999 by
ST. MARTIN'S PRESS, INC.,
Scholarly and Reference Division,
175 Fifth Avenue, New York, N.Y. 10010

ISBN 0-312-22568-7

Library of Congress Cataloging-in-Publication Data
Subregionalism and world order / edited by Glenn Hook and Ian Kearns.
  p. cm.
Includes bibliographical references and index.
ISBN 0-312-22568-7 (cloth)
1. Regionalism (International organization)   2. International
economic integration.   I. Hook, Glenn D.   II. Kearns, Ian.
KZ1273.S83   1999
337.1—dc21   99-26117
                                                                  CIP

Selection, editorial matter and Chapters 1 and 11 © Glenn Hook and Ian Kearns
Chapter 2 © Ian Kearns 1999
Chapter 10 © Glenn Hook 1999
Chapters 3–9 © Macmillan Press Ltd 1999

All rights reserved. No reproduction, copy or transmission of this publication may be made without written permission.

No paragraph of this publication may be reproduced, copied or transmitted save with written permission or in accordance with the provisions of the Copyright, Designs and Patents Act 1988, or under the terms of any licence permitting limited copying issued by the Copyright Licensing Agency, 90 Tottenham Court Road, London W1P 0LP.

Any person who does any unauthorised act in relation to this publication may be liable to criminal prosecution and civil claims for damages.

The authors have asserted their rights to be identified as the authors of this work in accordance with the Copyright, Designs and Patents Act 1988.

This book is printed on paper suitable for recycling and made from fully managed and sustained forest sources.

10 9 8 7 6 5 4 3 2 1
08 07 06 05 04 03 02 01 00 99

Printed and bound in Great Britain by
Antony Rowe Ltd, Chippenham, Wiltshire

Contents

Preface		vii
List of Key Abbreviations		viii
Notes on the Contributors		xi

1. Introduction: the Political Economy of Subregionalism and World Order

    Glenn Hook and Ian Kearns — 1

Part 1 Subregionalism in Europe, Central Asia and West Africa

   Introduction by Ian Kearns — 17

2. Subregionalism in Central Europe

    Ian Kearns — 21

3. The Black Sea Economic Cooperation Scheme

    Gerasimos Konidaris — 41

4. West African Subregionalism: the Case of the Economic Community of West African States

    Stephen Riley — 63

Part 2 Subregionalism in the Americas

   Introduction by Jean Grugel — 91

5. MERCOSUR: from Domestic Concerns to Regional Influence

    Paul Cammack — 95

6. The Association of Caribbean States

    Anthony Payne — 117

7. Going it Alone? The Chilean Strategy for Subregional Integration

    Jean Grugel — 139

Part 3 Subregionalism in East Asia

    Introduction by Glenn Hook   165

8   The Association of Southeast Asian Nations
    Dominic Kelly   169

9   Politics of Identities and the Making of the 'Greater China' Subregion in the Post-Cold War Era
    Ngai-Ling Sum   197

10   The East Asian Economic Caucus: a Case of Reactive Subregionalism?
    Glenn Hook   223

11   Conclusion: Subregionalism – an Assessment
    Ian Kearns and Glenn Hook   247

Index   259

Preface

This book has been written largely by members of the research cluster on regionalism based at the Political Economy Research Centre (PERC), the University of Sheffield. The group itself was formed in late 1993, along with several others, as PERC began its mission of addressing contemporary issues in political economy from an interdisciplinary perspective. Work on regionalism has been going on almost continuously since that time. The present volume should be seen by readers as a direct follow-on from our first volume, *Regionalism and World Order*, edited by Andrew Gamble and Anthony Payne, which was published by Macmillan in 1996. The broad approach used in both volumes has been the same, and the present work treats subregionalism as something which exists in the shadow of the former, larger regionalism. For *Subregionalism and World Order*, however, the editors also would like to offer special thanks to the writers from outside PERC, in particular to Paul Cammack and Stephen Riley, who agreed to contribute chapters to the book and thus also lend their considerable knowledge and expertise to the project. Their efforts have helped to complement and strengthen the knowledge of members of the PERC group. We also owe a particular debt of gratitude to the Toshiba International Foundation and the Chubu Electric Power Company for helping to fund our work. As a result of their generosity we were able to organize international symposia and benefit from the views of some of Japan's leading social scientists.

Ian Kearns
Glenn Hook

Sheffield

Sadly, there is one final comment that we must make here. Since completing the writing and editing of this book the editors have learned of the sad and untimely death of one of the contributors, Steve Riley. Steve made a valuable contribution not only to this volume but also, over many years, to our understanding of the troubled region of West Africa more generally. He will be greatly missed and we hope that in some modest respect at least, this volume can stand as a testament to Steve's commitment to his subject and to his longstanding desire to share his knowledge of it with the rest of us.

List of Key Abbreviations

ACS	Association of Caribbean States
AEC	African Economic Community
AFTA	ASEAN Free Trade Area
AMCHAM	US–Chile Chamber of Commerce
APEC	Asia-Pacific Economic Cooperation
ARATS	Association for Relation across the Taiwan Straits
ARENA	National Renovation Alliance
ARF	ASEAN Regional Forum
ASA	Association of Southeast Asia
ASEAN	Association of Southeast Asian Nations
ASEM	Asia–Europe Meetings
BRASS	Black Sea Region Association of Shipbuilders and Ship Repairers
BSEC	Black Sea Economic Cooperation Scheme
BSTD	Black Sea Trade and Development Bank
CACM	Central American Common Market
CARICOM	Caribbean Community
CARIFTA	Caribbean Free Trade Area
CBI	Caribbean Basin Initiative
CBTAG	Caribbean Basin Technical Advisory Group
CEAO	Communauté Economique de l'Afrique de l'Ouest
CEFTA	Central European Free Trade Area
CEI	Central European Initiative
CFA	Communauté Financière Africaine
CIEPLAN	Corporacion de Investigaciones Economicas para Latinoamerica
CITIC	Investment Company
C/LAA	Caribbean/Latin American Action
CMEA	Council of Mutual Economic Assistance
CPC	Federation of Producers and Traders
CSCE	Conference on Security and Cooperation in Europe
DPP	Democratic Progressive Party
EAEC	East Asian Economic Caucus
EAEG	East Asian Economic Group
EAI	Enterprise for the Americas Initiative
EC	European Community
ECLAC	Economic Commission for Latin America and the Caribbean

List of Key Abbreviations

ECOMOG	ECOWAS Monitoring Group
ECOWAS	Economic Community of West African States
EMU	European Monetary Union
EU	European Union
FDI	Foreign Direct Investment
FEDEFRUTA	Fruit Growers Federation
FTAA	Free Trade Area of the Americas
FYROM	Former Yugoslav Republic of Macedonia
G7	Group of Seven
G8	Group of Eight
GATT	General Agreement on Tariffs and Trade
GDP	Gross Domestic product
IBSC	International Black Sea Club
IPE	International Political Economy
ISI	Import Substituting Industrialization
ITC	International America and the Caribbean
MDB	Brazilian Democratic Movement
MERCOSUR	El Mercado Comun del Sur
MFN	Most Favoured Nation
MNC	Multinational Corporation
MOFERT	Ministry of Foreign Economic Relations and Trade
NAFTA	North American States
NATO	North Atlantic Treaty Organization
OAS	Organization of American States
OAU	Organization of African Unity
OECD	Organization for Economic Cooperation and Development
OECS	Organization for Eastern Caribbean States
OPEC	Organization of Petroleum Exporting Countries
PDC	Christian Democratic Party
PDS	Social Democratic Party
PLA	People's Liberation Army
PMC	Post-Ministerial Conference
PMDB	Party of the Brazilian Democratic Movement
PPD	Popular Democratic Party
PRC	People's Republic of China
PS	Socialist Party
PSD	Social Democratic Party
PSDB	Brazilian Social Democratic Party
PT	Worker's Party
PTB	Brazilian Labour party

RN	National Renovation Party
SAR	Special Administrative Region
SEANWFZ	Southeast Asia Nuclear Weapons Free Zone Treaty
SEATO	Southeast Asia Treaty Organization
SEF	Straits Exchange Foundation
SELA	Latin American Economic System
SEZ	Special Economic Zones
SFF	The Manufacturer's Association
SNA	The National Farmers Society
SONAMI	The National Mine Owners Society
SSA	Sub-Saharan Africa
TCA	Trade and Cooperation Agreement
TVE	Township-Village Enterprise
UDI	Democratic Independent Union
UDN	National Democratic Union
UEMOA	Economic and Monetary Union of West African States
UNECA	United Nations Economic Commission for Africa
UNIDO	United Nations Industrial Development Organization
WTO	World Trade Organization

Notes on the Contributors

Paul Cammack is Professor of Government at Manchester University.

Jean Grugel is Lecturer in Politics at the University of Sheffield.

Glenn Hook is Professor of Japanese Studies at the University of Sheffield.

Ian Kearns is Lecturer in Politics at the University of Sheffield.

Dominic Kelly is Research Fellow in International Relations at the Department of International Studies and Law, Coventry University.

Gerasimos Konidaris is a doctoral candidate in the Department of Politics, University of Sheffield.

Ngai-Ling Sum is Simon Research Fellow at the International Centre for Labour Studies, Manchester University and was formerly Alex Horsley Research Fellow in the Political Economy Research Centre at the University of Sheffield.

Anthony Payne is Professor of Politics at the University of Sheffield.

The late Stephen Riley was Reader in Politics at Staffordshire University.

1 Introduction: the Political Economy of Subregionalism and World Order

Glenn Hook and Ian Kearns

The purpose of this book is to fill a lacuna in the empirical literature and to contribute to the theoretical debate on regionalism by bringing together in one volume analyses of the various forms that subregionalism is taking in the emerging world order. It seeks to explain the origins, developments and essential features of the subregionalist projects promoted by a number of the weaker states in Europe and Africa, in the Americas, and in East Asia. As most theoretical work on the 'new' regionalism of the late 1980s and post-Cold War era draws on the regionalist projects promoted by the big powers, with empirical work on the European Union (EU) and to a lesser extent the Asia Pacific Economic Cooperation (APEC) at the heart of these endeavours, our analyses of subregionalist projects promoted by the weak also should contribute to deepening our theoretical understanding of the trend towards regionalism in the contemporary world. Thus, the reader hopefully will find *Subregionalism and World Order* provides both theoretical and empirical insights on subregionalism in the nascent world order.

The book starts from the premise that despite the large amount of literature now available on regionalism, something new can still be said by bringing together case studies of subregionalist projects in different parts of the world in a theoretically informed way. We aim to say something new in a dual sense: first, in terms of the case studies presented. The chapters discuss subregionalism in Europe – the Central European Free Trade Agreement (CEFTA) and the Black Sea Economic Cooperation Scheme (BSEC); Africa – the Economic Community of West African States (ECOWAS); the Americas – the Common Market of the South (MERCOSUR), the Association of Caribbean States (ACS), and Chile, which has sought subregional ties with both MERCOSUR and NAFTA; and East Asia –'Greater China', the Association of South East Asian Nations (ASEAN), and the East Asian Economic Caucus (EAEC). This is the first time a single volume has covered these varied subregionalist projects. Second, in terms of the theoretical approach adopted, while differences remain in emphasis, reflecting both the individual author's eclecticism as well as the nature of the subregionalist project under scrutiny, each case study, where relevant, takes account of politics, economy, ideas (culture), and security. As discussed in the Conclusion, this allows us to draw out similarities and differences in the subregionalist projects pursued by these weaker states.

The theoretical approach adopted here differs from a variety of other approaches to regionalism and subregionalism which can be found in the extant literature. A review of these approaches can be found in Andrew Hurrell, on which the following draws.[1] In the first place are systemic theories or approaches to regionalism. Classical realism or its more recent variant, neo-realism, seeks to explain why, given the anarchical and conflictual nature of the international system, states cooperate at the regional level. This approach seeks the answer to why the European Community (EC) emerged in the nature of the international system, paying attention to such things as power politics and the geopolitical environment of the Cold War, especially the pressure on the Europeans from the hegemonic United States and the threat to their security from the Soviet Union. From this perspective, regionalism can be understood as a means for states to enhance their bargaining power, balance a bigger power, or as a way to 'entrap' a more powerful state in a regional framework, as in the case of Germany in Europe.

The other main systemic approach pays attention to the interconnectedness of states in the international system, especially economic interdependence promoted by the new information technologies. Thus, much recent work examines regionalism in the context of globalization, which draws attention to the varied economic, political and social linkages amongst states. It is not always clear how globalization and regionalism are related: in some instances, globalization creates pressure towards establishing new institutional frameworks going beyond the state and the region. In others, it sets in motion pressures for regional cooperation, as regional answers to global problems can often prove easier to implement on the regional rather than the global level. Indeed, financial globalization is eroding the power of the state to control events, giving rise to the need for regional institutional frameworks to address issues of common concern.

In comparison with these systemic approaches, neo-functionalism, neo-liberal institutionalism, and constructivism all approach regionalism from the 'inside-out'. With a primary focus on Europe, neo-functionalists have emphasized how economic links and interdependence within a region over time lead to political cooperation due to 'spill-over'. Spill-over is seen to occur either functionally, due to the need to solve the technical problems arising through greater cooperation and integration, or politically, due to institution-building as part of the integration process. In this way, neo-functionalism posits the embedding of institutional frameworks for regional cooperation and 'spill-over' from one area to another, with a plurality of actors, such as multinational corporations and interest groups, involved in the process of region-building as well as states.

In contrast to neo-functionalism, neo-liberal institutionalism gives centre stage to the state, which is seen as a rational actor cooperating on a regional institutional level in order to cope with problems shared by other states in the region. Thus regional institutions seek to promote the common good of the member states by addressing issues of shared concern, and in so doing help to shape the interests, norms, and expectations of the states as members of the regional institution. In this way, neo-liberal institutionalism sees regional institutional frameworks emerging in response to the concrete needs of states to manage regional problems and as a means of helping to reduce the costs of strengthening intra-regional linkages, as in the case of economic transactions amongst regional states.

The third approach, constructivism, pays especial attention to the inter-subjective nature of regional groupings, where developing a regional identity or a shared sense of belonging is seen as an essential part of institutionalizing regional cooperation. Instead of taking a rationalist approach, with the state very much as a given, constructivism examines how the interests and identities of plural actors are constructed within the context of different histories, cultures and processes of interaction. It is in this sense a social more than an economic approach to regionalism, where the importance of the 'soft' rather than the 'hard' aspects of regional cooperation are regarded as central. The constructivist approach thus seeks to understand how the sharing of ideas, knowledge, and norms contribute to the emergence of regional cooperation and regional institutions.

Finally, three domestic-level approaches link regionalism to shared features of the domestic situation in members of the regional grouping, paying attention to state viability, regime type, and convergence of domestic policy preferences. Briefly, the first points to the importance of a viable state before regional cooperation can forge ahead. It is precisely in cases where the state is viable, legitimate, and its boundaries are largely uncontested, as in Europe, that regional cooperation has succeeded. In this sense, rather than regionalism being an alternative to the state, viable states are its prerequisite. As far as regime type is concerned, the discussion has revolved around the question of the extent to which democracy plays a role in promoting regionalism. The existence of democratic regimes can help to explain regionalism in certain parts of the world, as in Europe, but not in others, as in East Asia. The last approach highlights how regionalism can be explained by reference to the goal of maintaining specific domestic policies. From this perspective the EC can be understood not as an outgrowth of internationalist concerns, but rather as a way to maintain a particular type of social welfare state and economic policies based on Keynesianism.

As Hurrell points out, a variety of criticisms have been levelled at these different approaches – for instance, that neo-realism ignores the importance of domestic factors, that globalization and regionalism are difficult to untangle, or that neo-functionalism downplays the state's resistance to regional cooperation in the area of 'high politics' – but overall they have enhanced our general understanding of regionalism. Still, the need remains to try to draw together insights from both the 'hard' and the 'soft' approaches to regionalism, on the one hand, and from viewing regionalism as involving both state and non-state actors, on the other. As discussed below, new international political economy ('new IPE') seeks to overcome some of the divisions in these different approaches by creatively combining a number of their important insights. The contributors to this volume all draw on the new IPE in investigating their specific subregionalist project.

APPROACH

This approach was first put forward in *Regionalism and World Order* (Macmillan, 1996), jointly edited by Andrew Gamble and Anthony Payne. The present volume follows on from this earlier work, which was carried out by the research cluster on regionalism at the Political Economy Research Centre, The University of Sheffield. In that book we investigated the form regionalism was taking in the three core regions of the global political economy, the relationship between globalization, regionalization and world order, and the constraints imposed on states by Cold War structures and the opportunities emerging for states as a result of the Cold War's ending. By approaching regionalism and world order in such a way we sought to extract regionalism from the journalistic debate on whether a struggle was emerging in the post-Cold War era amongst three titanic economic blocs, which were seen by some to be set on an ineluctable course of competition and conflict if not war, and place it under the dispassionate light of social scientific scrutiny. We premised our discussion on a distinction between regionalism, a project promoted by states; and regionalization, a process driven forward by markets.

Empirically, we found that regionalism can be understood as a reaction to globalization and the development of regionalist projects elsewhere, with the European commitment to introducing the Single Market and the European Union acting as a spur to the development of NAFTA and APEC. Moreover, whether in Europe, the Americas, or East Asia, the talk of 'blocs' hardly matched the complex reality of regionalist projects and regionalization processes in the triadic cores of the global political economy. Rather than the 'closed' regionalism evinced by the image of 'fortress' Europe, the prattle about a 'defensive' US economic strategy leading to a closing of American markets, or 'Japan Incorporated' as the centre of a new East Asian economic behemoth, we found complex, overlapping linkages and interconnections amongst these different regional political economies to be the order of the day. Nor, at the other extreme, did we find regionalism as a statist project being swamped by the forces of globalization, with the world turning borderless as multinational corporations moved effortlessly and swiftly from one part of the world to another in pursuit of profit. Instead, in all three regions, the state remains a powerful force in shaping and responding to both regional and global forces, with the Cold War structures continuing to impose constraints on the behaviour of certain states, especially in the realm of security, and the end of the Cold War providing opportunities for others to pursue new forms of regional cooperation.

Theoretically, the work sought to demonstrate that, in investigating the trend towards regionalism within the post-Cold War world order, old talk of the 'hegemonic stability thesis' and the straitjacket of orthodox IPE needed to be complemented if not completely replaced by a new approach. 'New IPE' is the term given to a critical international political economy which balks at the positivist premise of mainstream IPE, rejects its 'problem-solving' agenda, and seeks a more nuanced understanding of the emergence of global and regional orders through analysis of the 'soft power' of ideas as well as the 'hard power' of material forces. This approach draws on the work of Robert Cox,[2] who has sought to introduce insights from Gramsci into IPE, particularly the need to take account of ideas and the consensual as well as coercive nature of power; and Susan Strange, who has drawn attention to four structures of power – production, knowledge (ideas), finance (credit) and security.[3] Thus, the new IPE seeks to analyse regionalism and world order by taking into account the four dimensions of politics, economy (finance, production), security and ideas (knowledge, culture), thereby drawing together both the 'hard' and the 'soft' aspects of power. It pays attention to the historical background against which new regionalism has emerged as well as to the social forces which now are seeking to promote specific regionalist projects. In this sense, 'new IPE' pays attention to both state and non-state actors in promoting regionalist and subregionalist projects.

LEVELS OF REGIONALISM

In our discussion so far we have not drawn a distinction between regionalism and subregionalism, although the present volume deals with the latter. The term 'subregionalism' has been adopted in order to distinguish the case studies presented here from the higher levels of regionalism promoted by the big powers and the lower levels of microregionalism ('sub-subregionalism' or, in certain cases, 'sub-state regionalism') promoted by national and subnational actors. By 'higher levels of regionalism' we refer to regionalist projects like the EU and APEC. Regionalism in this sense includes regionalist projects promoting preferential trading arrangements (the EU) as well as non-discriminatory practices (APEC). By 'lower levels of regionalism' we refer to the microregional projects like the so-called 'growth triangle' linking the national economy of Singapore with the subnational economies of Johor in Malaysia and Riau in Indonesia as well as the microregionalist projects aimed at tying together the subnational economies of more than one state, as in the case of the Japanese city of Niigata, which is promoting the Japan Sea Zone concept as a way to link the Japan Sea prefectures with parts of the Korean peninsula, the Russian Far East, and China.[4]

Of course, as 'higher' and 'lower' are metaphors used to create a sociospatial structure and understanding, 'region', 'subregion' and 'microregion' remain contested concepts, with the 'subregional' and 'microregional' levels as contested as the 'regional'. For the process of inclusion and exclusion within these overlapping levels of regionalism is a highly political process. By here using the term 'subregionalism' we wish to draw attention to the subregionalist projects promoted by the weaker states in the global political economy which are seeking to strengthen cooperation in a more circumscribed space than at the regional level. In this sense, subregionalist projects take on their significance within the context of the more embracing regional projects and identities promoted by the more powerful states.

The difficulty of coming to grips with subregions is that changes in the international system, as well as questions of identity, influence our understanding of their boundaries. In the Cold-War era, space as a source of identity was suffocated beneath the weight of bilateralism at the heart of the ideological confrontation between the East and the West. The regional groupings created in this environment tended to demarcate their boundaries with reference to ideology as well as spatiality, so that the widening and deepening of the EC took place within, rather than across, the East–West ideological divide. The power of the epistemological bifurcation of the world during the Cold-War era thus meant that the EC appeared as a 'regional' organization. Looking back from the perspective of the post-Cold War era and the expansion of the community to the East, however, it now appears as a subregionalist project. Again, as far as identity is concerned, in order for Finland to be embraced within the European regionalist project, Finnish policy-makers needed to generate a 'European' identity for Finland.[5] In this sense, moving the boundaries of the European identity to Scandinavia was as much a part of the process of region-building as was the inclusion of Finnish representatives in EU institutions.

In the case of East and Southeast Asia, in contrast, the bilateral pull towards the American heartland meant that ideology rather than contiguous spatial relations was central to the creation of regional organizations like the Southeast Asian Treaty Organization (SEATO), a 1954 US attempt to graft Pakistan onto a pro-American 'Southeast' Asian military organization composed additionally of Cambodia, Laos, the Philippines, Thailand, and South Vietnam, which soon became defunct. Ideology was also central to the Association of Southeast Asia (ASA), an anti-communist political alliance formed in 1961 by Thailand, Malaya and the Philippines. Likewise, despite a rhetorical commitment to political 'neutrality' and a more clear-cut thrust for capitalist-style economic development, during the Cold-War era ASEAN embraced anti-communist Indonesia and the Philippines within the new grouping, not contiguous communist regimes, and widened later only to include an independent Brunei due to a propinquity of interests. In this way, the Cold-War years saw the knife used to carve up 'Asia' into 'Southeast Asia' being honed with anti-communist ideology and developmental capitalism, not the spatial contiguity at the heart of the cartographer's view of 'Southeast Asia'.

With the end of the Cold War, the ideological boundaries of the East–West divide have given way to contestation over spatial demarcation. For if space is at the heart of attempts to build new regional and global orders, then state as well as non-state actors will seek to impute spatial relations with meaning as part of a contested socio-political process, thereby redefining their place in a world where the death knoll is sounding on the compass point mode of understanding, 'East–West, North–South.' At the regional level, we see the breakdown in the East–West political divide in eastern and central Europe's orientation to the West, redefining the boundaries of the European heartland. The breakdown in the North–South economic divide is clear from the attempts to embrace within subregional and regional groupings the economically developed as well as developing. In this way, the trends towards regionalization and regionalism in economy, politics, ideas, and security highlight the way in which boundaries are being contested and redrawn in the process of building new regional and global orders in the post-Cold War era.

STRUCTURE OF THE BOOK

What is this situation like on the subregional level? The case studies included in this volume are meant to shed light on this question by demonstrating how the weaker states in the international system have been coping with the globalization of political economy, the creation of regional institutions by the bigger powers, and the needs of development in the post-Cold War world. The book is divided into three sections, with an Introduction to each section in order to set the scene for the case studies which follow. Here we will provide an overview of some of the key issues raised in each chapter.

The first chapter by Ian Kearns examines CEFTA against the background of other attempts at multilateral cooperation in central and east-central Europe, such as the Alpe-Adria Working Community and the Central European Initiative. CEFTA seeks to promote economic cooperation, reduce tariffs, and has even promoted coordination on certain foreign policy and security issues. The proposal to pursue this project in the 1990s grew out of the longer-term process of defining 'central Europe', with intellectuals as well as politicians involved in this task. But the question still remains: to what extent is 'central Europe' united with a clear identity, or divided with the central European idea and identity contested? In addition to these internal questions, the relationship between CEFTA and the external environment also is addressed: has the difficulty in entering the EU stimulated subregional cooperation by central European states? In seeking answers to these and other questions, Kearns demonstrates how CEFTA has emerged, not out of any shared sense of identity or culture, but rather as a pragmatic response to the economic and security pressures from outside the subregion. In this sense, the 'hard' and the external, rather than the 'soft' and the internal, are crucial to explaining the emergence of this subregional project.

Gerasimos Konidaris addresses the BSEC in the next chapter. This subregional group took on life in 1992, with the aim of promoting closer cooperation, especially in the fields of economics and technology; ensuring peace and stability in the subregion; and deepening the economic integration of the Black Sea states into Europe and the world. Cooperation, which has been moving forward tentatively in a variety of areas – economy, technology, transportation, education, and so on – takes place within a formal institutional framework, including the Meeting of the Ministers of Foreign Affairs and a Permanent International Secretariat. Through these bodies the BSEC aims not only to strengthen cooperation amongst its members, but also to develop links with external multilateral organizations, especially the EU. So far, the ambitious schemes for cooperation have faced numerous hurdles due to the problems these economies face in their transition to Western-style market economies and pluralist democracies, bilateral tensions between some of the members, and their lack of competitiveness in the global economy. Still, the Black Sea states need the organization as a means to enhance their competitiveness and build links with other economies. As Konidaris concludes, for many of the former communist states, the possibility of entering the EU, while a strong desire, remains a distant possibility. In this sense, the BSEC offers them a way to develop much closer relations with the EU than could be expected by acting unilaterally.

The third chapter on ECOWAS by Stephen Riley takes us away from Europe in order to shed light on how developing African states, which maintain strong ties with the developed West, have been moving forward with subregional cooperation. Unlike the two previous chapters, which have taken shape in the post-Cold War 1990s, this is a subregional project seeking to promote economic integration with roots in the 1960s and 1970s, which has been revitalized in the 1990s. It is important as one of the four pillars upon which it is proposed to construct the African Economic Community in the twenty-first century. However, despite the original goal of promoting economic integration and this future orientation of ECOWAS, the organization in the 1990s has increasingly become concerned with issues of security, as in ECOWAS's military involvement in the conflicts in Liberia and Sierra Leone. In the face of long-term patterns of trade tying African economies to Europe, strengthening economic links within the subregion has proven difficult, especially given the heterogeneity of the subregional states. As mainly primary-commodity exporters, the states of this subregion are peripheral to the regional institutions now being strengthened in Europe, the Americas and Asia Pacific, with Nigerian oil being of greatest interest to outsiders. How successful has subregional cooperation been in reducing dependence on the developed countries? As Riley states, without Nigeria changing into a less rent-seeking democratic polity, and without subregional economic cooperation being strengthened, the greatest success of ECOWAS will remain as a 'political-cum-security organization'.

The three chapters in the next section focus on the Americas. It starts out with Paul Cammack's contribution on MERCOSUR, which highlights the degree to which subregionalist projects emerge in some states in response to domestic needs. The organization was launched in the 1980s and expanded in the 1990s, with a customs union and a common external tariff as concrete manifestations of the success of the enterprise so far. How does the success of MERCOSUR in terms of size and function relate to the major political changes affecting Latin America in this period, the emergence of democratic regimes in say Brazil and Argentina? Cammack's point is that the move to democracy has gone hand in hand with a domestic project to establish a neo-liberal economy: military regimes have been replaced by emerging liberal democracies and import-substitution by market-led answers to development questions. What is more, MERCOSUR is seeking increasingly to develop stronger economic links with the EU, rather than the United States. In the case of this subregionalist project, therefore, the goal of establishing a position in the global political economy is of less importance than imposing the type of social and political relations required for a liberal market economy to flourish. How domestic as well as international factors have influenced the emergence of MERCOSUR as a significant regional player is a particular point of concern to Cammack.

The fifth chapter by Anthony Payne discusses the ACS. The ACS has emerged in the 1990s as a subregionalist attempt by a group of Caribbean states to avoid possible marginalization in the increasingly competitive global political economy. It seeks to bring together a large and diverse group of countries in order to establish a new regional architecture meant to enhance the interests, especially economic interests, of the members through intergovernmental cooperation. The task of promoting these interests involves the Caribbean in carving out a new relationship with both the United States and NAFTA. It might also mean responding to the establishment of the Free Trade Area of the Americas which, despite Congress's denial of 'fast-track' authority to President Clinton in 1998, is supposed to come into being in 2005. Another major concern that the ACS has to address is security, which in the context of the Caribbean means more the threat to domestic political order and drug-trafficking than the guns, bombs and tanks at the centre of security concerns in many other parts of the world. How are countries more tied to the developed economies of the West than to each other, and threatened by financially strong drug cartels, able to enhance their common interests? As Payne points out, the weakness of the ACS in being able to deal with either the economic or security needs of the Caribbean is all too plain to see, given that this subregional group's role might well be reduced to the mundane tasks of promoting the three 'ts' – tourism, trade, and transportation – rather than the far more important task of dealing with the development and security issues that the Caribbean faces.

Jean Grugel's contribution next takes up the case of Chile, which has sought subregional links with both MERCOSUR and NAFTA. As the title, 'Going it Alone?' suggests, Chile is the only country in Latin America and the Caribbean set on a dual course of seeking association with two subregional groups. It is a member of APEC, too. How to account for this Chilean strategy is one of Grugel's main concerns. It is a strategy which, in comparison with other Latin American countries, has roots in the militarist era. For even under General Pinochet economic liberalization and the restructuring of capitalism went ahead in the 1970s and, again, from the mid-1980s onwards. As a result of these and other changes implemented after the move to democracy, Chile has emerged as an investor-friendly, market-led, export-oriented economy trading more with the EU and Japan than with the United States. Such an orientation, Grugel suggests, has resulted from a particular coalition of domestic forces, which brings together the state elite and exporters. It is this coalition which determined the subregionalist strategy pursued in respect of both MERCOSUR and NAFTA. In this way, the Chilean approach to subregionalism can be seen as an outgrowth of an elite commitment to liberalism – economically; politically, albeit in a limited sense; and in terms of the development model championed. That many Latin Americans regard the costs of this 'Chilean Way' to be too great, especially in the face of widespread income disparities and millions of poor, points to the underlying threat to security that this subregionalist strategy contains.

The final section of the book concentrates on three case studies from East Asia: Dominic Kelly on ASEAN, Ngai-Ling Sum on 'Greater China,' and Glenn Hook on EAEC. Kelly starts by examining the establishment of ASEAN in the 1960s and the reinvigorated role it has been playing in post-Cold War East Asia in order to shed light on the association as a process, with an identity and aspirations, not just as an organizational structure. Regional political leaders often promote the idea of 'Asian values' and an 'ASEAN way'. Can such 'soft power' be used to explain the cohesion of ASEAN? Or, are historical circumstances, especially the decline in American hegemony, more important? Unravelling the complex relationship between 'hard' and 'soft' power is a particular point of concern to Kelly. He is also keen to examine the concrete results of ASEAN. For in pursuing economic growth and development in order to try to promote peace and stability in the subregion, the political elite have remained strongly attached to the status quo and have been reluctant to democratize. This has meant that economic disparities and injustices remain widespread. In this sense, the 'ASEAN way' has not necessarily benefited the peoples of Southeast Asia. If the 'ASEAN way' has any validity, Kelly suggests that this is not so much to be found in the field of political economy, where poverty and authoritarianism remain at home and inequality and hierarchy remain internationally, but in the field of security. The role that the association has been playing in promoting security dialogue through the ASEAN Regional Forum thus points to the possibility that the 'ASEAN way' will be able to contribute to the resolution of conflicts in Southeast Asia.

The eighth chapter by Sum looks at a case of what could be called informal subregionalism, as 'Greater China' is more an identity than a subregional grouping in an organizational form. This identity has been promoted by the People's Republic of China on the basis of the vast array of economic and social linkages which have developed amongst Hong Kong, Taiwan and Southern China in the wake of the Chinese government's 'Open Door' policy. The Chinese networks at the heart of Greater China share what Sum calls an 'imagined community' of interests, which are realized in the global, regional, national and subnational contexts of these 'three Chinas'. Social forces utilize different discourses and discursive strategies in pursuing this task, but 'Greater China' involves the 'hard' as well as the 'soft' aspects of power. For the material base at the heart of this subregional project, which includes trade, investment and production systems, provides the wherewithal for a new political identity. Indeed, the 'Open Door' policy can be said to have created the political and economic environment for the subregion to emerge, as the southern provinces of China were thereby given the opportunity to pursue a more independent economic strategy in the wake of the commercial and financial reforms introduced. This change in Chinese policy also offered opportunities for Hong Kong and Taiwan to reformulate their trade and investment strategies towards the mainland. How the politics of identity have emerged in Hong Kong, Taiwan and China in the wake of the emergence of a complex web of overlapping economic and financial ties is a particular point of concern in Sum's chapter.

In the last chapter Hook takes up the case of Malaysian Prime Minister Mahathir's attempt to create an East Asian subregional grouping. At one level, the EAEC can be regarded as a reactive strategy pursued by a developing country seeking to ensure survival in the context of the widening and deepening of the European project and the launch of NAFTA and APEC. At another, however, it can be seen as a strategy to promote Malaysian interests in an East Asian political economy dominated by Japan, which plays the key role in investment, trade and production in the region. Within a broader, structural and historical context, however, EAEC appears as a mechanism by which a late-comer, developing nation is seeking to respond to the transformation of the global political economy. From this perspective, Malaysian policy elites have sought in subregionalism a means to respond to the wider structural changes brought about by globalization, regionalization, and the end of the Cold War. Despite the slow progress made in institutionalizing EAEC, the holding of intra-subregional meetings and the establishment of the Asia–Europe Meetings (ASEM) in 1996 provide venues for East Asian nations to promote their common interests, albeit without a declared EAEC identity. How EAEC has evolved in the face of resistance by the United States and Japan is a particular point of focus in this chapter.

In this way, the contributors to this volume seek to enhance our understanding of subregionalism in different parts of the world. They each highlight the essential features of a subregionalist project promoted by the weaker powers in the international system. Although not of one mind in respect of the reasons for the emergence of these subregionalist projects and the course they will take, with some paying more attention to domestic factors and others to international, each contribution nevertheless offers fresh empirical and theoretical insights into subregionalism and world order.

NOTES

1. A. Hurrell, 'Explaining the Resurgence of Regionalism in World Politics', *Review of International Studies*, 21, 4 (1995) pp. 331–58, and also A. Hurrell, 'Regionalism in Theoretical Perspective', in L. Fawcett and A. Hurrell (eds), *Regionalism in World Politics* (New York: Oxford University Press, 1995), pp. 37–73.
2. R. Cox with T. Sinclair, *Approaches to World Order* (Cambridge: Cambridge University Press, 1996), especially chapters 6 and 7.
3. S. Strange, *States and Markets: an Introduction to International Political Economy* (London: Pinter, 1994).
4. On the latter, see G. Hook, 'Japan and Subregionalism: Constructing the Japan Sea Economic Zone', *Kokusai Seiji* [Special 40th Anniversary Issue, Journal of the Japan Association of International Relations], 114 (1997), pp. 49–62.
5. Comment by P. Korhonen, international symposium on regionalism, University of Sheffield, February 1996.

Part 1
Subregionalism in Europe, Central Asia and West Africa

Introduction

by Ian Kearns

The chapters in this section of the book are somewhat unique in that they cover developments on more than one continent. Indeed, though partially European centred, the focus here is wide enough to include both Central Asia and parts of West Africa, too. Despite the fact that this obviously means a great deal of diversity among the subregions covered, one should also perhaps start out by noting some of the features which bind the three subregions of CEFTA, BSEC and ECOWAS together. All represent attempts at subregional cooperation within what might be termed the periphery of the world economy. All, to some extent at least, also reflect attempts to use cooperation as a tool for overcoming this peripheral status. The one political judgement which stands in common among all three, therefore, is the belief that the promotion of a greater economic regionalization within each subregion may hold the key to advancement. Naturally, the detail of each attempt at this greater regionalization has been different in each case and has also been complicated by other motives and divisions which are unique to each subregional grouping alone. The aim in this section of the book, then, is to sketch out and analyse those elements of similarity and of difference which together make up the contours of subregionalism in each of the areas mentioned.

In central Europe, the drive to subregional cooperation can be traced most obviously to the one facilitating factor which overrides all others, namely the end of the Cold War division of the continent and the subsequent opportunity for all states in the area to reinvent their international relations, at least to some degree. Subregionalism in this area has also been attempted in the context of an uneven process of state dissolution and new state formation and has been further set against the background of the political and economic shift of the eastern half of the continent away from communism. There is a sense, then, in which subregionalism has been part of a deeper and more far-reaching process of change and, more specifically, in which subregionalism is part of an attempt by the states involved to reintegrate their economies into the structures of global capitalism. This focus on economics, furthermore, has prevailed despite the fact that the original boost for cooperative efforts came from a fear that reactionary forces may once again take control in the Soviet Union during 1990 and 1991.

The same, in many ways, could also be said of subregional cooperation in the Black Sea area. Certainly the end of the Cold War and in particular the collapse of the Soviet Union has been critical. Many of the states involved in BSEC, some of which are again newly created, are also undergoing exactly the same post-communist transformations as those in central Europe. This is true whether one examines parts of the Balkan peninsula or the republics of the former Soviet Central Asia. There are also, however, some important differences in context between the Black Sea Scheme and CEFTA in central Europe. On the one hand, some of the cooperation which is developing in the former is a natural realignment, given that several of the central Asian republics are Turkic speaking. The closer relationship of these states with Turkey is to be expected now that the Soviet Union has ceased to exist. Indeed it is Turkey itself, perhaps, which is the central linchpin of this whole cooperative effort since, by virtue of both its history and its geography, it looks both to the east and to the west. Turkey undoubtedly wishes to be a power in the Balkans and in Europe more generally, as its applications to join the EU and its angry responses to EU rejection have shown. But Turkey is now also an Asian power and is seeking to prosper from the Soviet collapse.

On the other hand, BSEC also represents an attempt to manage a potentially very explosive and unstable security situation *within* the membership of the organization itself. Relations between Greece and Turkey and between Armenia and Azerbaijan, for example, are in a very precarious state. Military conflicts cannot be ruled out and the unspoken truth for many in the non-Turkic-speaking membership is that BSEC is a vehicle to tie in and limit Turkish ambitions. That said, however, one must not ignore the fact that BSEC is unique in that it ties parts of Europe and central Asia together in an attempt to boost levels of economic development for all. It also, by definition, is one of the only fora within which parts of south-eastern Europe have been labelled as developing territories. In a context where, for far too much of the time, the fallacious assumption has been made that south-eastern Europe is just like western Europe except for the alien impact of the intervention of communism, this could turn out to be an interesting and significant psychological realignment.

The clear inclusion in BSEC of states which cannot be described as anything but developing, gives this organization also something in common with ECOWAS, the case of subregional cooperation which has emerged in west Africa. ECOWAS originally developed out of, and still pays lip-service to, the notion of pan-African unity and self-reliance as a mechanism for throwing off the old colonial legacy. Like BSEC then, ECOWAS on paper at least has arisen in order to help overcome the daunting problems of poverty and underdevelopment which face the states involved. Also like BSEC, however, ECOWAS has increasingly over time become dominated in essence by security constraints. ECOMOG, the military arm of ECOWAS, while carrying out its major peace-making mission in war-torn Liberia and during its activities to restore democratic government in Sierra Leone, has often seemed like the only notable activity taking place at the subregional level at all.

Despite these similarities, though, there are also reasons as to why ECOWAS must be distinguished from the other efforts at subregional cooperation which are dealt with in this section of the book. In particular, it is important to note that ECOWAS has become dominated by the interests of the Nigerian regime, the major actor at the regional level, and by the latter's attempts to use ECOWAS for its own advancement. Subregionalism then, in the chapters which follow here, is characterized both as a response to hegemony in one case (CEFTA as a response to the EU) and as an element of hegemonic strategy (the case of Nigeria) in another.

2 Subregionalism in Central Europe

Ian Kearns

INTRODUCTION

One of the new and more notable features in the political economy of central Europe in recent years has been the increased trend towards greater cross-border cooperation. This has been evident, both in a whole string of bilateral free trade agreements signed by and between the governments of the region in the 1990s, and in the various efforts at *multilateral* cooperation which have emerged since 1989. The latter, multilateral component of the process, clearly the more important and potentially far more significant development of the two, has included the continuation and expansion of the Alpe-Adria Working Community, the creation of the Pentagonale Initiative (later renamed the Central European Initiative or CEI) and the creation of the Central European Free Trade Agreement (CEFTA). Of these, CEFTA, in particular, has been generating a serious sense of momentum, intent as it is on a further expansion in size and membership, and may yet become an arrangement covering over 150 million consumers, thus making it a free trade arrangement on a world scale.

Given this general context, it seems timely to discuss the nature and origins of the ongoing wave of increased cooperation. Is it, for example, a genuinely new development, or more of a return to pre-Cold War patterns of international relations? Is it likely to last and, if so, what significance should ultimately be attached to it? The present chapter seeks to engage with and discuss some of these issues and, in doing so, is organized into three principal sections. The first of these sketches, in brief the substance of the primary schemes of cooperation noted above, charts some of the agreements reached, and outlines the mechanisms of cooperation being employed. The second and longest section assesses the underlying motivations which initially gave rise to cooperation, balancing off in the process an assessment of the internal 'push' and external 'pull' factors which appear to have been relevant. This necessarily also includes some discussion of the historical and cultural context in which cooperation arose as well as reference to the economic and security cirumstances of the states involved. The third and last section, albeit briefly, evaluates the longer-term significance of the most substantial multilateral schemes, with special emphasis being placed on CEFTA and its possible future role.

Before engaging with any of this agenda, however, one small issue of detail requires clarification. As the discussion below will make clear, the meaning of labels, not to mention identities, such as 'central Europe' and 'east-central Europe', is heavily contested, often to serious political end. It is not proposed here to engage in any meaningful way in this debate but merely to employ a very simple division in geographical terms. 'Central Europe', on the one hand, is used to refer to Italy, Austria, Switzerland and all of the former communist states of eastern Europe outside the former Soviet Union; 'east central Europe' on the other hand, refers *only* to the former communist states of this region and is, therefore, only a part of the broader region of 'central Europe'. For the most part Germany, due to its size, influence, and history of domination of others in the area, is not included as a part of either. In essence, what this means is that the focus of the chapter is on countries such as Poland, Hungary and the Czech and Slovak Republics, either working together amongst themselves or in collaboration with others within the region and/or with others across the old East–West divide.

RECENT MULTILATERAL COOPERATION IN CENTRAL AND EAST-CENTRAL EUROPE

The first attempt at multilateral cooperation in central Europe in recent times, the Alpe-Adria Working Community, was established in 1978 on the initiative of the Venetian local government and was formed as a body designed to pull together various of the provinces and regions of central Europe. This province rather than state-driven initiative has since included the northern provinces of Italy, the southern provinces of Austria, the whole of Croatia, Slovenia, Bavaria, a number of Swiss cantons and various of the comitats in Hungary. The Alpe-Adria organization sought initially to play a bridging role between East and West, including within its ranks areas which had previously either been within the Habsburg Empire or, alternatively, had been heavily and directly influenced by it. More recently the body's role has been one of the pure facilitation and stimulation of contacts and exchanges among its members, both in the hope of promoting economic, technical and educational cooperation and with the aim of attracting structural funds from the European Union (EU). Primary areas of concern for Alpe-Adria have been 'trans-Alpine communications, harbour traffic, production and transportation of energy, agriculture, forest use, tourism and protection of the environment.'[1] In carrying out its tasks, it has also been aided by a small secretariat and by a conference of the Presidents of the various regions which meets annually to discuss progress. The whole organization, however, only makes recommendations on cooperation and offers general encouragement to the process through the exchange of information. It does not in and of itself, either through the use of resources or through the inclusion of binding agreements among its signatories, consist of concrete commitments to cooperate.

Alpe-Adria was added to, and perhaps slightly bettered, in April 1990 by a further, and this time explicitly post-Cold War development in the form of the Pentagonale Initiative. This emerged at the inter-governmental rather than the provincial government level and initially included Hungary, Yugoslavia, Czechoslovakia, Austria and Italy. It sought to develop cooperation amongst small and medium-sized businesses, to develop cross-border infrastructure in telecommunications and transport and to promote cultural cooperation.[2] Poland joined the grouping in December 1990, thus briefly making the Pentagonale the Hexagonale, but before the newly expanded body could have any real impact it became paralyzed by the slide of one of its members, namely Yugoslavia, into civil war. At a meeting in July 1992, a decision was taken (without Yugoslavia) to relaunch the organization under the new name of the Central European Initiative (CEI), and to bring in new members. The CEI now includes Poland, Hungary, the Czech Republic, Slovakia, Austria, Italy, Croatia, Slovenia, Bosnia-Hercegovina and Macedonia. Its underlying aim is 'to promote regional policy coordination and inter-governmental cooperation on a functional basis' but it also involves a significant broadening out of the original Pentagonale agenda to include both the issue of national minorities in eastern Europe and the potential coordination of Western aid to the region.[3]

In pursuit of this again, a small and loose organizational structure has been developed, this time including a secretariat based in the Austrian Foreign Ministry and a series of working groups designed to focus upon particular functional areas of cooperation. Of these, the working group on national minority issues works permanently, while the others meet only occasionally. There is also an annual conference of prime ministers and biannual meetings of member state's foreign ministers, while a CEI parliamentarians committee meets alongside these higher level gatherings. Despite all of this, and despite the widened aims and high level meetings which go to make up the CEI, however, practical cooperation on the ground is still really only to be encouraged and not explicitly developed by CEI committees. As with the Alpe-Adria Working Community, the CEI's main instrument of influence is the promotion of cooperation through information exchange rather than the generation of binding agreements. Consequently and again as with Alpe-Adria, few major cooperative successes have been registered.

It is in this sense more than in any other that the CEFTA Agreement stands out both in importance and impact over and above either the Alpe Adria or CEI initiatives. For CEFTA is a major international agreement, not a mere statement of aspiration, and contains a series of commitments made by its signatories. It is thus an achievement of a different qualitative order and of more depth than either Alpe-Adria or the CEI. It deserves consequently to be treated at greater length here. The CEFTA agreement, initially signed by the governments of Poland, Hungary and Czechoslovakia in December 1992, emerged out of a broad process of diplomatic engagement between the former communist states of east-central Europe in the period 1990–92.[4] Initially, this process amounted to no more than a stock-taking exercise in the new, post-1989 environment and, to the extent that cooperation had an immediate aim, this resided only in the effort to coordinate a withdrawal from both the Warsaw Pact and the Council of Mutual Economic Assistance (CMEA). Consequently, the opening summit of the 'three' in Bratislava in April 1990 contained little by way of a forward-looking agenda. By early 1991, however, this situation had changed quite dramatically. At a further meeting in Visegrad, the three agreed a declaration of intent to increase cooperation, began talks on some minor though long-standing border disputes between them, and embarked upon a series of bilateral negotiating processes. The Czechoslovak and Hungarian delegations, in particular, soon began serious work on discussing the rights of Hungarians living in Slovakia and the controversial issue of the Gamcikovo-Nagymaros Dam project.[5] While no major diplomatic breakthroughs were achieved immediately on this agenda, the talking in its broadest sense was significant since it represented the most sustained effort to recast relations within the east-central European region in over half a decade. Furthermore, by the time the three held their next summit meeting in Krakow in October 1991, the agenda had broadened out even further to include the first attempts to negotiate the free trade agreement which later became known as CEFTA. Though these trade talks initially failed, largely due to insecurity on the part of all three states with regard to the impact of open trade on their weak and uncompetitive economies, the governments of the region persisted. By May 1992, agreement was reached in principle at a summit meeting in Prague to form a CEFTA and the three then committed themselves to working out its details as a matter of priority. CEFTA in this context, then, became the first and main tangible product of the broader attempts to cooperate within central Europe and east-central Europe which have just been outlined.

The final CEFTA agreement, when signed, initially covered a joint population of 64 million people and a land area of some half a million square kilometres. A wide range of areas of cooperation were mentioned in the text including transport and highway infrastructure, telecommunications, tourism and retail trade. A further series of microregional projects to bind border regions more closely together, a direct network of enterprise-to-enterprise interactions and the establishment of bilateral foundations to promote cooperation in the areas of scientific, historical and cultural understanding, all also featured. The agreement even contained a commitment to increase cooperation on security matters. The core of the document, however, concerned the creation of the free trade area.[6] This was to be accomplished by 2001 and would be achieved through a phased process of reductions in tariffs. In the first phase, trade was to become free with effect from March 1993 in industrial raw materials and some industrial manufactures. Sulphur, copper and salt exports from Poland, together with aluminium and bauxite exports from Hungary would, for example, be affected, as would pharmaceuticals production and trade in agricultural machinery. In the second phase, tariffs would be gradually removed from trade in most industrial products by the end of 1996. This itself would come in three stages: a one-third reduction in tariffs from January 1995, another one-third reduction sometime in 1996 and the final third by 1 January 1997. Finally, in the third phase of reductions, tariffs would be removed in particularly sensitive areas such as the motor industry (especially sensitive in Poland), textile production (sensitive in the Czech Republic), and steel (sensitive in Hungary). These goods would be traded freely within eight years of the initial signing of the agreement, thus making CEFTA nearly complete. Any goods which were already being traded freely before the agreement was signed were to be left alone or included in the first phase to ensure that new tariffs were not imposed. The one exception to all of this was the agricultural sector which, treated as a separate category, faced only partial reductions in tariffs even by 2001. Import quotas on certain agricultural goods were also to be maintained. This reflected the importance of agriculture to all members of the group and, in particular, the Polish and Slovak desire to protect their large agricultural sectors from more efficient Hungarian competition.[7]

The arrangements represented an ambitious programme of cooperation and tariff reduction and the guiding principle throughout was that all reductions of tariffs would take place on a strictly symmetrical basis. Initially the arrangements would affect around 25 per cent of the total trade among the members but, within less than four years, the timetable suggested coverage of as much as 85 per cent of all trade between them. The levels of cooperation thus looked impressive. And they began to look more impressive still when viewed against the broader political climate within which CEFTA was signed. For the 1992 Prague summit not only contained the agreement in principle to create a CEFTA but also agreed that the three should submit a joint application for membership of the European Union, that NATO, and in particular the United States, should be encouraged to play a clear and unambiguous role in east-central Europe, and that the former Yugoslav republic of Macedonia (FYROM) should receive their collective recognition. This was significant because it represented the first signs of a coordinated approach to foreign policy. By the time the CEFTA agreement was actually being signed at the end of 1992, therefore, it appeared that the former communist countries of east-central Europe had embarked upon a broad and deep process, not just of trade liberalization but also of cooperation in other significant areas, attempting both to harmonize their dealings with important external actors, and to open up and to integrate their economies more closely with each other.

Events since the signing have, on the surface at least, done little to contradict this. The whole process of CEFTA's implementation was in fact speeded up. In July 1994, it was agreed that the target date of 2001 for the free-trade zone was not ambitious enough and tariff reductions were accelerated. In August of the following year, it was even decided to intensify the liberalization of trade in agricultural products. Furthermore, in January 1996 the signatories agreed to expand their membership and accepted the former Yugoslav republic of Slovenia into their ranks. They also decided that all those countries which had both signed an association agreement with the European Union and had accepted the General Agreement on Tariffs and Trade (GATT) could in principle be eligible for membership. Subsequently, Romania has become a signatory while Bulgaria, Lithuania and Estonia have asked to sign. CEFTA, then, appears both to be meeting its trade objectives and, as noted at the outset, to be growing in size.

No-one can doubt on the basis of this, that central and east-central Europe have seen at least some serious steps forward in attempts at cooperation in recent years. Perhaps the crucial issue in assessing the significance of all of this, however, is the extent to which the cooperative activity outlined rests and indeed builds upon a common sense of history, culture and identity among the peoples and states involved. If cooperation is a 'natural' development in these terms then it can be expected to last and to have significance in the post-Cold War world. If it is not, then we must both look elsewhere to understand it, and perhaps be more careful of over-emphasizing its importance in the longer term.

THE DRIVING FORCES BEHIND COOPERATION IN CENTRAL AND EAST-CENTRAL EUROPE

The efforts at cooperation outlined have, it must be said, developed against a longer-term backdrop of renewed interest in 'central Europe' as an idea. This itself is traceable back to the later years of the Cold War and to the early and mid-1980s in particular. The latter period, for example, saw both the re-emergence of the concept of 'Mitteleuropa' into German political discourse and the use of 'central Europe' as a rallying cry around which dissident intellectuals under communism could attempt to gather. Even within 'east-central Europe', however, what is meant by the term 'central Europe' has not always been clear. Some, such as Milan Kundera, basically define central Europe as a part of the West which was forced into the East: as a 'piece of the Latin West which has fallen under Russian domination and which lies geographically in the centre, culturally in the West and politically in the East'.[8] Further support for this view comes from Gyorgy Konrad, the leading Hungarian intellectual, who argues not only that central Europe is Western by inclination but also (and here is the real point of the argument) that the West is far superior to the east and, in particular, to Russia. As Garton Ash put it during a review of Konrad's work: 'We are to understand that what was truly Central European was always Western, rational, humanistic, democratic, sceptical and tolerant. The rest was east European, Russian or possibly German.'[9]

Popular though these views are today, especially among those in east-central Europe who see their utility in justifying a political move to the West, they are not without difficulties. The most obvious concerns the very selective history of both 'West' and 'East' which underpins them. In particular, the belief that Western influences on central European development have been unquestionably for the good, while manifestations of evil, such as communism, have been alien interventions from Russia, is nothing short of a myth. One might, for example, as Garton Ash has done, point out the stark and painful truth that 'if Kafka was a child of central (and therefore by implication, Western) Europe, so too was Adolf Hitler'.[10] Or, alternatively, adopt the line of Joseph Brodsky's attack on Kundera which notes that 'the political system that put Mr. Kundera out of commission [communism] is as much a product of Western rationalism as it is of Eastern emotional radicalism'.[11] Even if Konrad and Kundera's optimistic view of the West could be sustained, however, the history of central Europe itself would make it difficult for the region to meet the exacting requirements of Western membership. Many of the features of communism in central Europe during the Cold War years, for example, drew heavily upon much older traditions indigenous to the region. In particular, a commitment to superbureaucratic statism and formalistic legalism taken to incredible extremes were both prevalent in central Europe before 1914 and were features upon which the communist regimes could and did indeed build. In this sense, central Europe is not quite as clean as some would have us believe. At least part of its historical woes have been self-inflicted.

What this difficulty with the views of Kundera and Konrad also points to, of course, is the broader difficulty of making clear distinctions between what is 'West' and what is 'East'. It is precisely this difficulty which has led some, such as Hyde-Price, to attempt to argue that central Europe, and in particular east-central Europe, is neither in the West nor the East but is instead, something quite unique. This, he points out, stems from the fact that east-central Europe has been influenced 'by all the great experiences of European civilisation – feudalism, medieval Christian universalism, the Renaissance, the Reformation, the Counter-Reformation and the Enlightenment',[12] and yet at the same time has been penetrated by Byzantium, Orthodox Christianity, the Ottoman Turks and the values of western Asia. States like Poland and Hungary have, as a consequence, been exposed to the cultural influences of the East as well as the West: the centre has been a 'melting pot' and, as a result, is a distinct region in its own right.

Again the problem with this uniqueness and a problem of direct relevance to our discussion here, is that when one tries to identify its boundaries and content, the very notion of it seems to slip away. Timothy Garton Ash, for example, went *In Search of Central Europe* in the late 1980s only to find that definitions of it were either 'absurdly reductionsist or invincibly vague'.[13] He was driven to this conclusion by the ongoing and increasingly interactive reflections of leading east-central European intellectuals who attempted to define their common bond only in spiritual rather than in geographical or political terms. The shared history of subjugation, the common sense of irony developed in the face of it, and a civic culture promoted by their own activities increasingly were, for them, the essence of central Europe and of what it meant to be a central European.

Garton Ash's complaints were insightful, however, not only for illustrating the vagueness of even spiritual notions of central Europe but also because they drew attention to the reason for shifting the discussion on to the spiritual level in the first place, namely the inability to agree on definitions of the region in any other way. 'Like Europe itself', he comments, 'no one can quite agree where central Europe begins or ends. Germans naturally locate the centre of Central Europe in Berlin; Austrians in Vienna. Tomas Masaryk defined it as a peculiar zone of small nations extending from the North Cape to Cape Matapan and therefore including Laplanders, Swedes, Norwegians and Danes, Finns, Estonians, Letts, Lithuanians, Poles, Lusatians, Czechs and Slovaks, Magyars, Serbo-Croats and Slovenes, Romanians, Bulgars, Albanians, Turks and Greeks – but no Germans or Austrians!'[14] Crucially, disagreements such as these over the geography have always also, of course, been disputes about politics. Indeed the very definition of the region noted by Masaryk above was developed explicitly to oppose the German usage of the term Mitteleuropa, which was thought by him to be a concept devised to satisfy German imperialist ambitions.[15] The various attempts, then, to pin down the essence of a Central European identity or geography have been both unconvincing and have ended in, or perhaps more properly reflected, division rather than unity.

The perception of a region divided rather than united is also reinforced by an examination of the central European historical context. On the one hand, it is true to say that there have been plenty of discussions and attempts at regional cooperation over the years. As Bakos pointed out, for example, 'plans for a Danube Confederation date back as far as the late eighteenth century, when the stability of the Habsburg empire was beginning to be eroded and national movements were starting to emerge'.[16] In this particular instance the cooperation envisaged included the creation of a federative structure for territorial units within the empire, with each unit enjoying equal status. This plan was later modified by the Hungarian national leader, Kossuth, in the middle of the nineteenth century into a call for a Danube Alliance. The Alliance was to have included Hungary, Transylvania, Romania, Croatia and various south Slavic provinces and would have operated via a federative parliament and council with powers to collectively manage trade, customs, transport, currency, foreign affairs and defence. The Alliance overall was also to guarantee the cultural rights of all those who populated its territories. Even the early part of the twentieth century saw the desire to cooperate continue amongst some in central Europe. In the 1920s and 1930s, for example, the idea took on the form of the 'Little Entente' which included Czechoslovakia, Yugoslavia and

Romania while in the brief period between the end of the Second World War and the onset of the Cold War a number of cooperative developments took place. These included a Czechoslovak–Polish agreement to increase economic cooperation and a Yugoslav–Bulgarian declaration of a customs union. Hungary and Yugoslavia also signed a bauxite–aluminium agreement and the logic of all the bilateral arrangements was that the next stage might produce some multilateral cooperative effort.

On each of the historical occasions mentioned, however, the notion of multilateral cooperation fell foul of the stronger forces opposing it. In the eighteenth and nineteenth centuries it was not, for example, only or even primarily the power of Habsburg rule which kept the subject peoples apart but was also more particularly the impact of Hungarian nationalism. The case of Slovakia provides an interesting illustrative example of this. Initially, Slovak intellectuals attempting to generate a sense of Slovak national identity had preferred to deal more closely with their Czech counterparts. 'The first Slovak literary forms had been deliberately based on the peasant dialects of western Slovakia, as being nearer to Czech even though these forms were strange to the bulk of Slovak peasants. Only later, and to save them from Magyarization, did the Slovak intellectuals adopt instead the dialect of central Slovakia. Thus, it was to meet the Magyar danger that the Slovaks were driven to become a *separate* nation.'[17] Elsewhere too, in the same period, Serbs and Romanians were fighting vicious wars against the Hungarians in order to avoid falling under their control, while Czech nationalists struggled fiercely for control of Bohemia, though, in the latter case, characteristically using only peaceful means. In short, in the eighteenth and nineteenth centuries, the central Europeans were clearly divided amongst themselves, often violently, and it was this basic division which destroyed the dream of multilateral collaboration.

Divisions also continued into the twentieth century, with inter-war central and eastern Europe being characterized by a string of border disputes and territorial claims. The only form of regional cooperation to emerge in this period, the 'Little Entente' of Czechoslovakia, Yugoslavia and Romania, was once again explicitly directed against Hungary.[18] Such difficulties in relations were only partly frozen with the even less preferable domination of the region first by Germany from the late 1930s, and then later by the Soviet Union from the 1940s on. Consequently, in this context, it seems reasonable to argue that the current wave of multilateral cooperation in Central Europe needs to be viewed as a break with, and not a natural outgrowth of, the region's troubled past. The break though, even now, has not been clean as any discussion of the contemporary political climate, particularly in east-central Europe, makes abundantly clear.

On the one hand, several elements within the post-communist political scene have proved supportive to the notion of multilateral cooperation. Again, some of the leading intellectuals and activists of the communist years have been important here. Vaclav Havel, in particular, formerly the President of Czechoslovakia and at the time of writing the President of the Czech Republic, has been a great and influential supporter of central European integration. Even he at times, however, has been outshone by others such as former Polish President, Lech Walesa, who allowed his enthusiasm for the CEFTA project in particular to become so great as to warrant talk of an 'EC-2' in the heart of Europe.[19] More broadly, the impact, in some quarters, of the rhetoric of the 'West in East' thesis also has generated support for a new approach to the international relations of the region. If being like the West is argued to mean being democratic and of a more tolerant persuasion then this in turn helps to steer attention away from past conflicts. Furthermore, when seen in combination with the advent of globalization and increasing economic interdependence, it also has the potential to allow some modification to traditional notions of sovereignty in a direction supportive to multilateralism.

Nevertheless, politically, the situation since the collapse of communism has not been as smooth as the discussion so far might suggest. The Czech and Slovak split which took effect from 1993, for instance, did little to signal a greater willingness to organize at the subregional level. Indeed, as Shumaker pointed out, this split 'brought to the forefront both Czech and Slovak leaders who were willing to sacrifice the federal state for greater autonomy'.[20] Furthermore, Slovak independence itself placed minorities questions higher on the political agenda since it sat uneasily with the wishes of the large Hungarian minority living in Slovakia. Their language and cultural rights have since formed a core element of debate in independent Slovak politics and, not surprisingly, have periodically generated tensions between Bratislava and Budapest. This particular bilateral relationship has also suffered through much of the 1990s from the effects of the long-running dispute over the Gamcikovo-Nagymaros hydro-electric power project. Agreed by both the Czechoslovak and Hungarian communist leaderships in the 1980s, this project involved a serious diversion of the flow of the Danube near the Slovak–Hungarian border. However, since the collapse of communism, the new Hungarian political elite has rejected the plan on regional environmental grounds and, in particular, has expressed concerns over its possible impact on local agricultural land and farming communities. The Slovak authorities, meanwhile, in need of domestic energy sources upon which they can rely, have continued to support the development.

Efforts at subregional cooperation have also been hampered by the fact that cooperation has had its powerful enemies as well as its influential friends. In 1990, for example, the then Czechoslovak Prime Minister and acclaimed reformer, Vaclav Klaus, could not have made his views on the matter any clearer. 'I do not think,' he said 'that there has ever been a real need to become involved in these activities.'[21] Later, and once cooperation was underway, the line of Klaus's attack was modified slightly, though his tone again did little to mask his preference for dealing directly with the West. 'I am against,' he said, 'all institutionalisation of central European cooperation and that is why I protest against this. I think that we should cooperate with Hungary, Poland, Austria and with other states, but this does not mean we should employ thousands of bureaucrats, give them high salaries and send them abroad all the time to attend meetings of this or that commission.'[22] The combination of Klaus's lack of interest and of Slovak feuding with the Hungarians over the Gamcikovo-Nagymaros project and the minority rights issue, then, indicates clearly that moves to central European cooperation have faced their difficulties and opponents. One further complicating factor in the 1990s, namely the level of ongoing political instability in Poland, has also added to the problems. Numerous Polish coalition governments have taken office only to collapse just a few months later during this period and not until the adoption of electoral laws limiting the number of parties securing seats in the Sejm did the situation improve. Even then, for most of the period during which central European and, in particular, CEFTA cooperation was being negotiated the improvement was only marginal. Effectively, there has been a sense throughout in which the reliability of Polish involvement in regional agreements of any kind could not be taken for granted.

When we turn our attention to an exploration of the economic conditions within which CEFTA in particular was created a further series of problems immediately becomes apparent. On the one hand, east-central Europe as a whole was both in deep recession and massively dependent on the West for capital at the start of the post-communist period. This was essentially the product of over four decades of communism since the latter's legacy was characterized by inefficient enterprises, poor quality management, technological backwardness, chronic underinvestment and serious budgetary crises. Failed attempts to overcome such difficulties by borrowing from the West had also generated an additional problem of large foreign debts for post-communist governments to face. On the other hand, the Soviet use of the Council of Mutual Economic Assistance (CMEA) during the communist years to tie east-European states primarily to the Soviet economy meant that in practice east Europeans were trading very little with each other at the end of the 1980s.

Such a general picture created a number of very specific barriers when it came to increasing economic cooperation within the east-central European region. First, it meant that the chances of CEFTA facilitating an increase in intra-regional investment on a significant scale were virtually non-existent from the outset. New investments after 1989 would certainly tie the east European economies more to the West and still not to each other. Given the links between patterns of investment and trade, this also meant an increase in intra-regional trade shares as a percentage of total trade would be difficult to achieve. In short, establishing a strong intra-regional orientation to investment and trade would require a contradiction of the historical and structural economic realities within which cooperation developed. Dependence upon the West also generated a further significant problem of relevance here, too; namely, that as sites for Western investment, potential partners of the West in trade, and possible early entrants to the European Union, the countries of former communist east-central Europe sat as competitors with one another at least as much as they were potential collaborators. This obviously was not the ideal context within which to engage in serious attempts at subregional cooperation.

It seems, then, that several elements in the historical, economic and political context within which central and east-central European cooperation emerged and has had to operate have not lent themselves to the easy development of the project. And yet, as noted earlier, the most impressive aspect of recent developments, CEFTA, has not only emerged but has also over time developed quickly in relation to its goals and has even expanded its membership. What, then, are the forces driving this particular instance of subregional cooperation?

Three issues need to be dealt with in providing an answer to this question. The first of these concerns the nature of the emerging relationship between east-central Europe and the European Union during the period 1990–92. As noted earlier, the former communist states were extremely keen to attach themselves to the West and appeared to believe that the response they would encounter from its institutions would be nothing but positive. Throughout 1990 and into 1991, however, as the EU response to the collapse of communism took shape, it suggested far from an early accession by the east-central European states to EU membership, a very narrow defence of EU self-interest even at the east-central European's expense. This appears clear from the string of bilateral Trade and Cooperation Agreements (TCAs) signed between the EU and east-central European states in 1990. Theoretically at least, these initial steps toward trade liberalization were to provide the latter with some access to new hard currency export markets and also went on to contain several clauses which suggested stimulation of a wide range of areas of economic cooperation including joint-venture promotion, scientific and technical cooperation, licensing agreements and exchanges of personnel. The problem with the TCAs, however, was highlighted by those aspects of trade which the agreements did not cover. The agreements excepted products covered by the Treaty establishing the European Coal and Steel Community, existing international Agreements concerning trade in textile products, and specific agreements or arrangements covering agricultural products.[23] In short, in the sectors in which a liberalization of trade relations with the EU would have benefited central European exporters the most, the EU kept up a protectionist guard to placate vested interests within the Union itself. While this hypocritical position was being defended, the EU also continued to preach to the governments of Poland, Hungary and Czechoslovakia on the merits of cooperation among themselves. Indeed, EU officials began to stress that the ability of the former communist states to engage in such cooperation would be viewed as a sign of the maturity required for eventual EU membership.[24] Furthermore, as time passed and the relationship began to shift from the bilateral TCAs to the negotiation of so-called 'Europe' agreements, it remained evident throughout that although the 'Europe' agreements would be broader than their predecessors (including, for example, cooperation on energy, environmental and infrastructure projects), the EU policy with regard to its most sensitive markets would not quickly change.

Faced with this position and noting that the EU was giving very similar draft agreements to each central European state negotiating a Europe Agreement, some in the east-central European leadership 'came to the conclusion that co-operation and exchange of information improved the bargaining position of each candidate for association, even though the negotiations were held separately'.[25] The actual negotiation of Europe Agreements, therefore, 'provided one of the major impetuses for an increased level of regional cooperation'.[26] The later failure, ultimately, to secure any greater concessions from the EU in the agreements signed, only served to magnify the lesson which it seemed necessary for the east-central Europeans to learn, namely that to some extent they were on their own. As dreams of their quick membership of the EU faded, therefore, the need for more regional cooperation among themselves came ever more clearly into view.

This realization was further reinforced in its effects by two other features of the immediate post-1989 international climate. One of these, the impact of the decision to switch intra-CMEA trade to a hard currency basis with effect from 1 January 1991, had very serious and damaging effects on levels of east-central European trade and ultimately meant that the economies of the region were struggling even more than the regional political leadership had originally anticipated. The other, the uncertainty and worsening security situation in the former Soviet Union, persuaded the east-central Europeans that their historic condition of vulnerability to great power interference had not disappeared and that their security position was best faced collectively. Each of these is important and is dealt with in more detail below.

The switch to hard currency intra-CMEA trading contributed to a major cycle of trade reorientation for the east-central European region. For the former communist economies suffering under the legacy of underinvestment and backward technology, using hard currency to buy less advanced and lower quality goods from one another was simply less attractive than trading with the West. In a short space of time, therefore, inter-enterprise trade within the former CMEA countries and, in particular, trade with the former Soviet Union, began to collapse. Even when taking into account the fact that between 1989 and mid-1992 the east-central European countries of Poland, Hungary and the Czech and Slovak Federative Republic enjoyed as much as 50 per cent increases in their levels of exports to the OECD countries, the sheer scale of the CMEA market collapse actually meant that in 1991 and 1992 all of these states were in fact trading and, in particular, exporting less in absolute terms than they had been in 1989. When added to a series of domestic difficulties with economic transformation, principally linked to the strategies for macro-economic stabilization and restructuring through privatization, this obviously exacerbated the post-communist recession already being experienced.

Once the seriousness of the position was realized, again, in practical terms the effect was to lend support to official efforts to stimulate levels of trade within the east-central European region itself. 'The Visegrad countries hoped that the CEFTA agreement would accelerate regional economic recovery through a revival of multi-lateral trade.'[27] That a need to *liberalize* trade in particular should come to the fore in this context has been further explained by Bakos.[28] According to him, the CEFTA agreement and in particular the trade liberalizing component of it, can at least partly be explained by the conditions of trade between Poland, Hungary and Czechoslovakia in 1990 and 1991. At around this time, steps towards some trade liberalization between the three had already begun to appear as a part of the broader economic transformation. The problem was that the first step to liberalization consisted only in the abolition of import regulations which had previously facilitated centralized control of trade.

The abolitions did not come alongside an immediate switch to real market prices and in the context of more open trade at artificial prices a series of distortions began to appear. This in its turn quickly produced protective measures. Poland, for example, increased import tariffs in August 1991 and this was followed by another customs increase in January 1992. Czechoslovakia also increased tariffs in January 1992 in such a way as to increase duties on Hungarian imports and especially to protect against Hungarian agricultural imports. East-central European cooperation in general, and CEFTA in particular then, were elements of a response both to the broader economic gloom, and to the very specific problem of growing protectionism in the immediate post-communist period.

Finally, in this treatment of the major external factors driving the cooperative process, the need to respond to a new, post-Cold War security situation must be revisited. The initial fear here for the Poles, Hungarians, Czechs and Slovaks, of course, was that there might be a challenge both to Gorbachev's leadership of the Soviet Union, and a reversal of the earlier Soviet decision to allow the east European satellite states to go their own way. This fear was not altogether without foundation since in the period 1990–91 Gorbachev clearly was increasingly in trouble. Many of the Soviet republics were attempting to extend their freedom and even secede from the Union, the Soviet economy was in deep trouble with no promise of improvement in the near future, and some 'hard-line' elements both within the communist party and the military were beginning to make themselves heard. What no-one could be sure of at the time was how many there were and how influential these disaffected groups might be.

For the east-central European states, the obvious and best solution to this problem was to join NATO. As with the response of the EU, however, NATO moved quickly to dissolve any notion that rapid membership of the organization might be achieved. In the immediate post-1989 period, the primary concern of NATO policy makers was to avoid doing anything in eastern and central Europe which might alienate the Soviet defence establishment, almost all of which was worried about a possible eastern expansion of the NATO alliance. Consequently, in between the end of the Cold War and the formation of CEFTA, only two sets of agreements, namely the Hand of Friendship Agreements of July 1990 and the Statements on Dialogue, Partnership and Cooperation of December 1991 were signed between NATO and the states of east-central Europe.[29] These agreements had in common the feature that they neither offered a security guarantee to east-central European states nor contained any notion of a timetable for NATO membership. In short, welcome though they were, they did little to remove the sense of exposure to threat prevalent across the region, and thus, fear of a renewed Soviet aggression and of instability more broadly, became a crucial driving force underpinning the move to CEFTA.

That this was the case is increasingly evident when one reconsiders the pattern of diplomacy leading up to CEFTA. It will be recalled, for example, that the first serious signs of cooperation in east-central Europe came at the Visegrad Summit in early 1991. This followed closely on the heels of a crackdown on nationalists in the Baltic republics by Soviet security forces in the winter of 1990/91. The first mention of a possible CEFTA, which came just a few months later at the October 1991 Krakow Summit, also itself followed only weeks after the failed coup attempt in Moscow in August of the same year. It further came as the hard-headed EU response to central European transformation was being made ever clearer during the negotiation of the 'Europe Agreements'. President Havel of Czechoslovakia was therefore able to credit the Krakow summit with 'co-ordinating the three's approach to the EU, solving or at least profitably addressing problems in their bilateral relations, and drawing the three together during "moments of crisis"'.[30] By May of 1992, a context of continued uncertainty in Moscow, ongoing war in Yugoslavia, and the failure of the international community to do anything effective about it produced the Prague Summit, which sketched out the principles of what would later that year become the CEFTA Agreement. The pattern of response to external pressures is therefore clear.

CEFTA AND THE FUTURE

The above analysis suggests then, that CEFTA, as the primary achievement to date of central and east-central European cooperation, has been a response to the economic and security pressures of the external environment rather than a natural reflection of shared identities, histories and cultures. Cooperation developed either to meet perceived external threats or to compensate for less than adequate external assistance in the process of post-communist transformation. In this, at least, CEFTA has not been altogether different from the CEI since the latter was also motivated by a sense of possible external threat, though this time from German economic power rather than from Soviet military might. That such a finding seems warranted, however, clearly also means that developments in the external environment will hold the key to CEFTA's, and for that matter the CEI's, future. Already, since the days of CEFTA's emergence, crucial relationships between east-central Europe, the EU and NATO have moved on.

Less than a year after CEFTA came into operation, the EU laid out the conditions to be met by any aspiring member states from the 'east-central European' region. These so called 'Copenhagen conditions' included, amongst other things, the ability of applicant states to take on the obligations of membership and also the existence and stability on their territory of democratic institutions and of a functioning market economy.[31] Such sweeping and rather vague requirements were later firmed up and clarified at the Essen Summit of EU leaders in 1994 where, as part of what became known as the EU's 'pre-accession' strategy, a 'Single Market White Paper' was developed. This represented a more formal attempt to prepare east-central European states for the detailed legal requirements of membership. By the end of 1997, and following a period to allow for adjustment to these requirements, EU leaders finally declared their readiness to enter membership negotiations with a first wave of east-central European candidates: these were the Czech Republic, Estonia, Hungary, Poland and Slovenia.

NATO also expanded its role and relationship within the region in this period. In 1994, for example, the pan-European Partnership for Peace Initiative was launched, which included facilities for military consultations and joint exercises among its signatories. This took cooperation beyond earlier agreements though it still fell far short of the goal of NATO membership from the east-central European perspective. Later that same year, however, NATO leaders at their summit meeting in Brussels moved further and confirmed in principle that they would welcome some enlargement of the organization to the east when the time was right. Details of a planned enlargement were finally announced at the Madrid Summit in 1997 and, as with the promise of EU expansion, it quickly became evident that this would take in only a few of the east-central European states: invitations this time being offered to the Czech Republic, Hungary and Poland. Running alongside the plans for this modest expansion was also a policy of greater engagement on security matters with Russia. Russian involvement in the Partnership for Peace Initiative had been secured in June 1994 and a real breakthrough, if NATO propaganda is to be believed, came in May 1997 with the signing in Paris of the Founding Act on Mutual Relations, Cooperation and Security between Russia and NATO. This effectively sought to bind Russia and NATO into a new partnership of trust and into an unprecedented level of cooperation which would span conflict prevention, joint peacekeeping operations and the free exchange of information on strategy, defence policy, and military doctrines.

Despite all of these changes in the broader scene, the important point for the purposes of this discussion remains the fact that such developments have not, even when viewed collectively, amounted to a fundamental transformation of the context which originally gave rise to efforts at subregional cooperation in east-central Europe. Membership of the EU remains at least 5–7 years away even for the front runners from the former communist world and, of course, is still nothing more than a distant dream for the rest. And despite the announced expansion of NATO, and the improved NATO–Russia relationship, the fears of instability and future conflict in the region will not disappear easily. One of the big question marks, for example, remains the future direction of Russia, especially in the post-Yeltsin period.

The need for cooperation, then, has not disappeared, but nor has cooperation itself been so succesful as to mean reduced dependence on the West. It would hardly be an exaggeration to say that all of the states of east-central Europe place their relations with the West in a higher category of importance than they do their relations with each other. Nowhere is this more evident or more obvious than in economics. Despite all of the progress on implementing CEFTA, the big trend of recent years has been the massive reorientation of trade away from former Soviet markets and towards the West. Intra-CEFTA trade amounts to only around 10 per cent of the total trade of the signatories of the agreement and even this figure is boosted somewhat by the disproportionately high levels of trade within the former Czechoslovakia. The impressive steps to free trade contained within the CEFTA agreement, therefore, affect only a small part of the region's overall economic activity. In these conditions, the importance of cooperation in central and east-central Europe cannot and should not be overstressed. Membership of all of the central-European entities discussed is useful at the margins and useful as a display of commitment to multilateralism but it remains a means and not an end – the end itself remains attachment to the European Union and NATO.

NOTES

1. A. Hyde-Price, *The International Politics of East Central Europe* (Manchester: Manchester University Press, 1996), p. 110.
2. See A. Reisch, 'The Central European Initiative: to be or not to be?', *RFE/RL Research Report*, 2, 34, 27 August 1993, pp. 30–37.
3. A. Hyde-Price, *The International Politics of East Central Europe* pp. 113–14.
4. D. Shumaker, 'The Origins and Development of Central European Cooperation: 1989–1992', *East European Quarterly*, 27, 3, September 1993, pp. 351–73.

5. J. Galambos, 'Political Aspects of an Environmental Conflict: the Case of the Gamcikovo-Nagymaros Dam System', in J. Kakonen, (ed.) *Perspectives on Environmental Conflict and International Relations* (London: Pinter, 1992), pp. 72–95.
6. See G. Bakos, 'After COMECON: a Free Trade Area in Central Europe?', *Europe-Asia Studies*, 45, 6, 1993, pp. 1025–44.
7. Ibid.
8. M. Kundera, 'The Tragedy of Central Europe', *New York Review of Books*, 26 April 1984, pp. 33–8.
9. See T. Garton Ash, *The Uses of Adversity* (Cambridge: Granta, 1989), pp 161–91.
10. Ibid., p. 166.
11. Quoted in ibid., p. 166.
12. A. Hyde-Price, *The International Politics of East Central Europe*, op. cit., p. 50.
13. See T. Garton Ash, *The Uses of Adversity*, op. cit., p. 169.
14. Ibid., p. 167
15. T.G. Masaryk was the President of Czechoslovakia for most of the interwar years.
16. See G. Bakos, 'After COMECON...', op. cit., p. 1025.
17. A.J.P. Taylor, *The Habsburg Monarchy, 1809-1918*, p. 125 (1948)
18. See A. Hyde-Price, *The International Politics of East-Central Europe*, op. cit., pp. 80–81.
19. See ibid., pp.108–33.
20. See Shumaker, 'The Origins and Development...', op. cit., p. 356.
21. Ibid., pp. 356–7.
22. Quoted in ibid., p. 369.
23. See T. Verheijen, 'The EC and Romania and Bulgaria: Stuck between Visegrad and Minsk', unpublished paper presented to conference of British International Studies Association, Swansea, 1992.
24. See A. Inotai and M. Sass, 'Economic Integration of the Visegrad Countries', *Eastern European Economics*, 1994, Nov.–Dec., pp. 6–28.
25. R. Stawarska, 'Poland's Association with the EEC', *Polish Western Affairs*, 33, January, 1994, p. 81.
26. L. K. Metcalf, 'Regional Integration? The Impact of CEFTA on Bilateral Trade Levels', paper presented to the International Studies Association Annual Conference, Toronto, Canada, March 1997, p. 5.
27. A. Hyde-Price, *The International Politics of East-Central Europe*, op. cit., p. 126.
28. G. Bakos, 'After COMECON...', op. cit.
29. J. Simon, 'Does Eastern Europe Belong in NATO?', *Orbis*, 37, 1, 1993, pp. 21–35.
30. D. Shumaker, 'The Origins and Development...', op. cit., p. 362.
31. H. Grabbe and K. Hughes, 'Redefining the European Union: Eastward Enlargement', *RIIA Briefing Paper*, 36, May 1997.

3 The Black Sea Economic Cooperation Scheme

Gerasimos Konidaris

Since ancient times the Black Sea area has been a cradle of different civilizations, a crossroads between Europe and Asia, which brought together peoples of diverse ethnic, cultural and religious origins. Throughout the centuries, the area has been well known for the linked economic and other relations of its peoples. These relations however, have not always been without problems. Periods of peace and stability, have often been followed by lengthy periods of conflict and war.

Today, on the eve of the new millennium – and following the end of the Cold War era and the changes that this has brought to the international environment – the countries of the region are facing new challenges with respect to their relations with each other and with the world around them. During the last few years – and in order to meet these new challenges – the countries of the Black Sea area are attempting to develop closer relationships among themselves, by putting forward common projects of an economic, technical, cultural and scientific nature.

To that end, on 25 June 1992, 11 countries of the region signed, in Istanbul, an agreement for the creation of the Black Sea Economic Cooperation Scheme (BSEC). In this chapter, we will seek to examine the nature and structure of this scheme, the type and level of the cooperation involved, and the objectives of its creation, given the particularities of the area and the problematic bilateral relations among some of the member states.

THE CREATION, PRINCIPLES AND OBJECTIVES OF THE BSEC

The thought of creating a Black Sea Economic area was conceived in January 1990 and it is primarily attributed to the Turkish diplomat Suekrue Elekdag.[1] Turkish President Turgut Ozal endorsed the idea and agreed to put it forward to the governments of Romania, Bulgaria and the Soviet Union. Consequently, in a preparatory meeting held in Ankara in December 1990, the deputy foreign ministers of the above-mentioned states agreed in principle to the establishment of an economic cooperation zone in the Black Sea and to the preparation of the necessary documents that would outline the nature of such cooperation. To that end, experts from these countries met on three occasions in Sofia, Bucharest and Moscow the following year.

It was during the third meeting in Moscow in July 1991, that the deputy foreign ministers agreed to adopt the documents of agreement and to put them forward for signature by the heads of state and government of the countries involved. However, political developments, primarily involving the disintegration of the Soviet Union, resulted in the postponement of their signature. The Black Sea Economic Cooperation Scheme was only finally established therefore at a summit meeting held in Istanbul on 25 June 1992. Two documents, the 'Summit Declaration on Black Sea Economic Cooperation' and the 'Bosphorus Statement' were signed by 11 countries including: the countries which emerged after the break-up of the Soviet Union (Armenia, Azerbaijan, Georgia, Moldova, Ukraine and the Russian Federation), Romania, Bulgaria, Albania, Greece and Turkey. The BSEC therefore, incorporated – apart from the littoral Black Sea states and countries from the Caucasus and the Balkans, a Scheme representing a geographical territory of around 20 million square kilometres and a population of nearly 330 million people.

According to the charter documents, the principles governing the BSEC are based on those of the Helsinki Final Act, the resolution of the CSCE follow-up documents, the Paris Charter for a new Europe, and other universally recognized principles of international law, as well as values, like democracy, rule of law, respect of human rights, and prosperity through economic liberty.[2]

The BSEC has three basic objectives as expressed in the 'Summit Declaration'. The first is the achievement of closer cooperation among the member states (and any other interested country) through the signing of bilateral and multilateral agreements, in order '... to foster their economic, technological and social progress, and to encourage free enterprise'.[3] The second objective, is '... to ensure that the Black Sea becomes a sea of peace, stability and prosperity, striving to promote friendly and good-neighbourly relations'.[4] And the third is economic cooperation to help the establishment '... of a Europe-wide economic area, as well as ... the achievement of a higher degree of integration of the Participating States into the world economy'.[5] In this context, the BSEC is considered by the heads of state and government of the countries involved, as '... an effort that would facilitate the processes and structures of European integration'.[6]

With respect to these BSEC objectives, two points should immediately be made clear. Firstly, concerning the economic cooperation of the member states, Article 10 of the 'Summit Declaration' notes that it will gradually take place by taking into consideration the interests of these countries, and particularly the economic problems that the countries of the ex-communist bloc are facing in their transformation into market economies. Secondly, the BSEC is regarded by the signatories not as an alternative scheme to other regional and international organizations and groups, but as a complementary endeavour in promoting European and international cooperation. This is depicted in Article 7 of the 'Summit Declaration' which states that the economic cooperation of the BSEC's members will be developed in a way '... not contravening their obligations and not preventing the promotion of the relations of the Participating States with third parties, including international organizations as well as the EC'.[7]

Article 18 of the 'Summit Declaration'[8] notes that other countries can also join the BSEC, provided that they are adopting the principles and aims of the scheme and the existing member states approve of their entry. Organizations, enterprises and firms, regional and international economic and financial institutions, can also participate in the realization of projects of common interest. Consequently, up to mid-1997, observer status in the proceedings of the BSEC had been given to the countries of Austria, Egypt, Israel, Italy, Poland, the Slovak Republic and Tunisia, and also two regional organizations, the BSEC Business Council and the International Black Sea Club. In April 1996, the Federal Republic of Yugoslavia and the Former Yugoslav Republic of Macedonia (FYROM), applied officially for full membership in the BSEC, while Jordan, Croatia, Slovenia, Cyprus, Kazakhstan and Bosnia-Hercegovina have applied for observer status. At the time of writing, no decision had yet been made with regard to these applications.

FIELDS OF COOPERATION

The areas of cooperation among the BSEC member states – as these have been outlined in the 'Summit Declaration' – are, *inter alia*, the following:

- transport and communications, including their infrastructure
- information
- exchange of economic and commercial information, including statistics
- standardization and certification of products
- energy
- mining and processing of mineral raw materials
- tourism

- agriculture and agro-industries
- veterinary and sanitary protection
- health care and pharmaceuticals
- science and technology.[9]

Obviously then, economic and trade cooperation in these areas are seen as central for the development of the BSEC.[10] The member states have decided to base their collaboration on these areas, respecting the principles of the World Trade Organization (WTO). Towards that end, it is the stated belief of the BSEC countries, that the harmonization of their national trade legislation would be a step in the right direction. Furthermore, the BSEC emphasis has been on the improvement of the business environment by promoting contacts among the enterprises and firms of the member states, as well as their collaboration with the private sector of other third countries. In relation to this, the 'Summit Declaration'[11] asks the participating states to facilitate the entry, stay and free movement of businessmen in their national territories; to support small and medium-sized enterprises; to promote the gradual lifting of economic, trade, and other obstacles (without contravening their obligations towards third countries); to create and support the proper environment for investment, capital flows and industrial collaboration (mainly by avoiding double taxation and protecting investments); and promoting cooperation in free economic zones. At their meeting in Moscow on 25 October 1996, the heads of State and governments of the BSEC countries, agreed to put forward a plan for the creation of a free-trade area in the region, a process which was officially adopted at the Special Meeting of the Ministers of Foreign Affairs (SMMFA) in Istanbul, on 7 February 1997, with the 'Declaration of Intent for the Establishment of the BSEC Free Trade Area'.

The improvement of the transportation system, has also been identified by the BSEC countries as a priority issue. The lack of an adequate and modern transport network in many of these countries, presents a serious barrier to their economic development. So, in their High Level Meeting in Bucharest on 30 June 1995, the member states emphasized the importance of improving 'the effectiveness and safety of the system of transportation ... in the Black Sea Region with a view to developing the national infrastructures with appropriate links to the Trans-European Transport System'.[12] BSEC bodies have discussed projects for the creation, expansion or improvement of railroads, highways and ports, which will efficiently connect the participating countries with each other, the neighbouring regions and other third countries. Two of these proposed projects would create a corridor connecting the Baltic Sea with the centre of Russia, the Azov and Black Seas, and link the Adriatic and the Black Seas.

Maps have been drawn up depicting the preferred transport links in the region up to the year 2005, which will be the basis for future proposed projects. During 1996, the BSEC started discussing possible collaboration on transport issues with the Central European Initiative (CEI) and others. With respect to maritime transportation in particular, the BSEC is also interested in putting forward bilateral shipping agreements among its members, on the basis of free and fair competition. It also aims to promote collaboration in the shipbuilding and ship-repairing sectors, in the form of joint ventures, with the involvement of both the member states and other third parties. For that purpose, the BSEC agreements build on the work of the Black Sea Region Association of Shipbuilders and Ship Repairers (BRASS), which was created in 1993. Another BSEC plan is the conversion of defence shipbuilding units of various member states into civilian ones. The Directorate General on Transport of the European Commission has also started to cooperate with the BSEC in identifying and solving practical problems existing in the ports of the participating countries.

Telecommunications is another area in which most BSEC states lack an effective, modern infrastructure. In the course of the last few years the BSEC countries have therefore put forward various joint-venture projects in order to improve communications among each other, and with the rest of the world. The first of these projects became operational on 31 October 1996 via a submarine fibre optical cable system of 3200km, connecting Italy, Turkey, Ukraine and Russia, with landing points in Palermo, Istanbul, Odessa and Novorossijsk. A second project is near completion which will connect Turkey, Bulgaria, Romania and Moldova.

In a study conducted in 1995, it was also shown that in relation to energy, the needs of the BSEC states are very diverse. Some countries, like the Russian Federation and Azerbaijan, have huge quantities of energy resources at their disposal, while others suffer from a lack of them. Nevertheless, the Group of Experts on Electrical Networks of the BSEC proposed the connection of the regional power systems of the member states, a system which will also be designed for exporting energy to neighbouring countries and regions. The BSEC has also started a project called 'Interconnected Regional Power Systems' with the aim of providing short- and medium-term solutions to the energy shortages of some participating states. Emphasis has further been given to the creation of new energy policies among the BSEC countries and to the development of a common electricity market in the region. A 'Memorandum on Cooperation' in the electric power industry has been signed by eight of the member states.

Complex environmental problems also exist in the Black Sea region; problems which have also attracted the interest of various international environmental bodies.[13] The BSEC has therefore emphasized the importance of collective action in confronting the environmental degradation of the region and, among other things, has proposed the creation of a special fund for the financing of environmental projects and closer cooperation of the Black Sea littoral states within a Special coordinating body. A BSEC–UNIDO Workshop for promoting cleaner production in the region is also in the making. The ministers for the environment of the six littoral states of the area agreed in Istanbul in October 1996, on the adoption of the 'Black Sea Strategic Action Plan', which will be implemented alongside the other two environmental projects of the region, the Bucharest Convention and the Odessa Declaration.

The BSEC member states also adopted in 1996 a 'Plan of Action' for the protection and improvement of their citizens' health. Among the proposed measures of this plan were: people's protection from infectious diseases and drug addiction; support of vulnerable groups in their societies; creation of a comprehensive list of the companies producing pharmaceuticals and medical equipment in the BSEC states; harmonization of legislation governing the medical and insurance firms of the region; and the improvement of the sewage and drinking water systems.

Cooperation has also spread to the area of law enforcement and, in particular, has involved closer cooperation of the member states in the fight against organized crime, terrorism, drug trafficking, smuggling of weapons and radioactive materials, and illegal migration. The successful control of these illegal activities would, it is believed, contribute positively to the economic development of the region.

The final areas of note include tourism, the exchange of information and statistics, and cooperation in the academic field. Despite the great potential for the Black Sea area to become an important tourist centre, the region is rather backward in this respect. Among the reasons for this underdevelopment to date are the lack of a proper tourist infrastructure and adequate services in most of the member states, and the low income of the native populations. Taking these weaknesses into consideration and having in mind the key role tourism can have in the economic progress of the region, the BSEC has since 1994 proposed various measures for the boosting of the tourist industry among the countries of the area and third countries. Within the region, the BSEC proposed to the participating states cooperation in the simplification and harmonization of the procedures in issuing passports, visas and other travelling documents (without contravening their obligations towards third parties). Concerning the movement from other countries, proposals to date have included increased investment in the tourist industry, promotion of health, hunting and fishing tourism, ecotourism, winter and yacht tourism, advertisement of the tourist opportunities of the BSEC region in tourism trade fairs outside the area and, creation of a regional travel agency, and hotel chain.

In the field of information and statistical data, the member states have established, within the State Institute of Statistics of Turkey, the BSEC Coordination Centre for the Exchange of Statistical Data and Economic Information. This will provide any interested party with information about the economic, trade, investment and other activities in the region. This Centre is in its initial phase and much further cooperation is needed to make it a reliable and comprehensive data source.

Academic cooperation in science and technology was also agreed in 1994. The basic aim of such collaboration is the use of the achievements of the academic communities of the BSEC states for the support and promotion of joint projects in the region. To that end, an International Centre for Black Sea Studies will eventually operate in Athens, which will also be studying ways of strengthening the economic relations within the Black Sea area, and fostering the contacts among the academic research centres of the region. The Centre further aspires to become a joining link between the BSEC, the European Union (EU), and the rest of the world. The Centre will produce interdisciplinary and policy-oriented research projects on economic, technical, social and other issues, aiming at a better understanding of the Black Sea region and its place in the wider world. The first BSEC Conference of representatives of the academic communities, took place in Athens on 9–11 December 1996. The participants agreed on the importance that such meetings could have for realization of the BSEC's objectives.

STRUCTURE

The areas of interest under the BSEC heading then are wide-ranging and many. To support such a diverse range of cooperative efforts, it was felt that BSEC needed some form of institutionalization. In the early days of its creation the BSEC was conceived by its member states, as a scheme having a very simple organizational structure, sufficient to give it the necessary flexibility for the effective accomplishment of its targets.[14] Without ruling out the possibility of establishing some permanent or *ad hoc* organs in the future, the only organ contemplated by the participant states at that stage was the Meeting of the Ministers of Foreign Affairs. No headquarters or secretariat was planned. In the course of time, however, and as new priorities entered the BSEC agenda, the structure of the scheme became bigger, and by 1997 comprised six main components: the Meeting of the Ministers of Foreign Affairs; The Subsidiary Bodies; the Permanent International Secretariat; the Parliamentary Assembly; the Business Council; and the Black Sea Trade and Development Bank.[15]

The Meeting of the Ministers of Foreign Affairs (MMFA) is the regular decision-making organ of the BSEC scheme. It convenes every six months and the presidency of the BSEC rotates among the member states in alphabetical order. There is a Chairman-in-Office, who also changes at the end of each presidency. In April 1995, a Troika System was adopted, which operates under the Chairman-in-Office. The decisions in the MMFA are taken either in unanimity or by a two-thirds majority, depending on the importance of the subject under consideration. From the BSEC's foundation to the end of 1996, there were eight meetings of the MMFA, and one more special meeting in February 1997.

The Subsidiary Bodies (SBs) – that is to say, the Working Groups (WGs) and the Groups of Experts (GEs) – are the instruments responsible for the elaboration of the various BSEC projects. The SBs correspond more or less to the numerous areas of cooperation of the BSEC scheme outlined above and they are composed of officials, technocrats and experts from the participating states. In their regular meetings the SBs discuss ways of implementing joint projects or forming new ones, and they forward their recommendations to the MMFA for consideration or approval. By the end of 1996, the SBs had met more than 80 times.

The Permanent International Secretariat (PERMIS) started functioning in full only on 10 March 1994 and its headquarters are in Istanbul. Its work is basically administrative, namely the preparation and distribution of the agenda for every MMFA to the member states; the support on administrative and secretarial issues in the BSEC meetings; the distribution of the BSEC's documents to every interested party; the organization and maintenance of the archives and documents of the scheme, and so on. The PERMIS is administered by a Secretary General, four Deputy Secretary Generals and a Project Coordinator as assistant to the Secretary General. All these posts are for a three-year period and the people appointed are chosen by the MMFA.

The Parliamentary Assembly (PABSEC) was created in June 1993. It is based in Istanbul and representatives from ten BSEC states (except Bulgaria), participate in its proceedings. The PABSEC is considered the instrument which gives a legal basis to the various forms of collaboration of the member states. The issues concerning the PABSEC are elaborated by three committees: Economic, Commercial, Technological and Environmental Affairs; Legal and Political Affairs; Cultural, Educational and Social Affairs. The reports and recommendations of these committees are then submitted to the General Assembly for further action. The work of the PABSEC is carried out by the International Secretariat under the supervision of the Secretary General. Parliamentary seats to the member states are allocated according to the size of their population. Therefore, countries with a population from 1 to 5 million (Albania, Armenia, Moldova) have 4 seats, from 5 to 10 million (Georgia, Azerbaijan, Bulgaria) have 5 seats, from 10 to 20 million (Greece) have 6 seats, from 20 to 50 million (Romania) have 7 seats, from 50 to 100 million (Ukraine, Turkey) have 9 seats, while those with more than 100 million (Russian Federation) have 12 seats. The representatives are not directly elected, but they are nominees from the national parliaments. The heads of state or governments of the participating states decided at their meeting in Moscow in October 1996, to think up a method for facilitating better cooperation between the PABSEC and the rest of the BSEC organs. It is not yet clear what the powers and jurisdiction of the Assembly will be.

The importance which the BSEC scheme gives to the business sector is clear from the creation of the BSEC Council, later renamed BSEC Business Council. The basic aims of the Business Council are, firstly, to promote private and public investment projects and secondly to develop closer relationships among the national business councils of the member states. It is governed by a Board of Directors and the Chairmanship of this Board rotates every six months. There is also a Secretariat which is based in Istanbul and is headed by the Secretary General. The first BSEC Business Forum took place in Bucharest in April 1996, with the participation of a large number of businessmen and other representatives from the private sector, the member states as well as third parties.

The idea for the establishment of a Black Sea Bank was originally embodied in the Istanbul 'Summit Declaration' of 1992 (Article 16) and is considered by many as the key element for a further successful development of the BSEC scheme. Named the Black Sea Trade and Development Bank (BSTD Bank) and headquartered in Thessaloniki, Greece, this bank is set to become the principal financial institution of the BSEC. The objectives of the Bank are the financing of joint projects in the region, the support of intraregional trade in manufactured goods, the provision of financial means for the member states, the evaluation of the commercial viability of proposed projects, and so on. The initial authorized capital of the Bank is 1 billion SDR (about $1.5 billion) and the participation rate of the member states in the share capital is: Greece, the Russian Federation, and Turkey 16 per cent each; Romania, Ukraine and Bulgaria, 13.5 per cent each; Azerbaijan, Armenia, Georgia, Moldova and

Albania, 2 per cent each. At the time of writing, the Business Plan of the BSTD had been completed,[16] and the conditions for the establishment of the Bank[17] had been fulfilled by all the participating states, except for Albania, Azerbaijan and Ukraine. Out of the countries that had fulfilled these conditions, however, only Greece, the Russian Federation and Turkey had deposited their respective paid-in capital to the European Bank for Reconstruction and Development, which has been appointed by the member states as the depository of the capital payments. Romania was expected to deposit its share next, while Armenia, Moldova and Georgia had been offered soft loans by Greece in order to facilitate the payment of their shares. These delays in payments meant that the inauguration of the BSTD Bank, scheduled to take place in Thessaloniki in May 1997, had to be postponed. Nevertheless, if it can be capitalized and made to work, it represents a further and potentially very significant aspect of the BSEC Scheme.

Before we conclude our discussion on the structure of the BSEC, we must also mention in passing the International Black Sea Club (IBSC), an initiative which brings together the municipalities of various important cities of the region. The IBSC is a non-profit organization with the status of a juridical person, aiming to promote the collaboration among the private sectors of the cities of the area, and to stimulate cooperation on ecological, cultural and other issues. The IBSC meets every six months. As yet, it is early days to evaluate its impact but, particularly in regard to tackling the problems of the large urbanized centres in the region, it may eventually play an influential role.

BSEC AND THE REST OF THE WORLD

As mentioned earlier, the BSEC is viewed by the participating states as a scheme open to any interested party which commits itself to the principles and objectives of the Istanbul 'Summit Declaration'. Apart from the possibility for third countries to become full members, the BSEC also offers the opportunity of closer collaboration with countries and international organizations on the basis of observer status.[18] Emphasis is also being given by the BSEC to the development of links between itself and other regional and international organizations and to relationships which can provide the BSEC with the necessary knowledge and experience to help in its efforts to improve the well-being of the peoples of the region. The main objectives of external contacts are therefore considered to be: better understanding of the activities and workings of other relevant international organizations; promotion of the BSEC objectives at an international level and exploration of possible support at the level of expertise and/or financial assistance to BSEC projects. Among the regional and international organizations with which the BSEC wants to establish closer relationships are the EU, the Organization for Security and Cooperation in Europe, the United Nations Economic Commission for Europe, the Council of Europe, the League of Arab States, the Economic Cooperation Organization, the Council of the Baltic Sea States, the Central European Initiative and the Euro-Mediterranean Initiative. Closer cooperation with the EU in particular is a stated priority for the BSEC.

AN EVALUATION OF THE BSEC'S ACHIEVEMENTS AND PROSPECTS

From the discussion so far on the principles, aims, fields of cooperation and structure of the BSEC, it is obvious that this scheme represents a rather ambitious effort on the part of its founding members to strengthen their economic potential and improve their economic performance, both at a regional and at an international level. The areas of proposed cooperation are many and of great importance for the economic development of the region. Despite the existence of so many plans and the expansion and improvement of its organization, however, it seems safe to say that the BSEC still has a long way to go. It is true that the economic and trade relations among some of the participating states have increased since the establishment of the scheme. Turkey for example, after 1992, has increased its exports to the BSEC countries from 3 to 57 per cent and its imports from 15 to 59 per cent.[19] It is also true that, in some areas of cooperation, like telecommunications and the BSTD Bank, progress is being, or about to be, made. In general terms though, the closer collaboration of the various proposed projects is well behind the levels which would make the BSEC a really important cooperation scheme. Most of the plans in the BSEC's agenda are still at the level of 'consideration', 'preparation', and 'consultation'.[20] Many of the SBs – which are to play a decisive role in the elaboration and promotion of the BSEC projects – malfunction or they hardly convene. By 1997 for instance, the WG on Banking and Finance had not met for more than three years, the WG on Cooperation in Science and Technology had not met for two years and the WG on Tourism met for the first and last time in October 1994.[21] This lack of dynamism of the BSEC scheme has been recognized by many of the heads of state or governments of the participating states and by the BSEC's institutions, all of which have stressed the necessity of improving the situation.[22] One might say, of course, that the BSEC is a new scheme, which needs time for its full development. It can also be said, however, that most of the member states have no great experience in the workings of such an economic initiative, as they have come out of nearly half a century of communist rule. The experiences of recent history have, therefore, clearly presented problems and barriers to deepening cooperation. One is that the whole BSEC process has been initiated in a period when most of the participating states are facing serious economic, political and social problems related to their transition into market economies and Western-type pluralistic democracies. Another is the existence of tense bilateral relations between some of the member states, tensions which in some cases are long-standing and have deteriorated occasionally into bitter conflicts. Further difficulties can also be related to the nature of the contemporary global economy. The BSEC has to face an extremely competitive economic environment, and other economic schemes and organizations with strong infrastructures and substantial experience with the functioning of world markets. Each of these areas needs to be discussed in turn.

As members of the former communist bloc, nine out of the eleven BSEC states are facing the grave consequences of transforming their command economies into market ones.[23] They have to confront high inflation, big budget deficits, external debts, high unemployment, problems with the convertibility of their currencies, and so on (for some indicative economic figures see Tables 3.1, 3.2, 3.3). They have also to privatize state-owned enterprises, to improve the infrastructure of their industries, transportation and communication systems, to increase productivity while reducing production costs, to introduce modern-style management and increase their competitiveness in world markets. At the same time they have to introduce and implement new legislation which will govern the new economic activities within their own national territories and their contacts with the rest of the world. This epigrammatic enumeration of some of the challenges that these countries have to meet for achieving their economic transition, represents of course a colossal struggle on their part, a struggle which has already started and which will go on for a long time to come. The difficult processes of change are having a great impact on the living conditions of the native peoples, who have to face unemployment, insecurity about their future prospects, decreases in the services offered by the state (health care, education, pensions and so on), and in many cases extreme poverty. Social unrest and political instability are often the direct results of such a situation and the countries in question are called upon to find the right balance between pursuing economic reforms and avoiding social dissatisfaction. As noted earlier, Article 10 of the Istanbul 'Summit Declaration' points out that the proposed economic cooperation in BSEC will only be gradually promoted and will take direct account of the economic conditions and other concerns of the BSEC States. What this means in practical terms, is that the member states are invited to participate in BSEC cooperation only to an extent that will not further jeopardize the preservation of their already fragile internal equilibrium. Thoughtful though this may be, however, one has to ask how beneficial such an attitude can really be for the future development of the BSEC scheme, or to put it another way, how retarding for its advancement the existing economic recession of the participating states will be, if we consider the enormous amounts of money needed for the promotion of the various projects involved. The example of the BSTD Bank is informative here. This Bank represents one of the areas of cooperation where satisfactory progress has been achieved (at least at an organizational level) and as noted is the instrument which will hopefully be the main financial institution of the scheme. Despite its importance, the Bank has not yet come into practical operation, because eight of the member states can not afford to pay their necessary share of the capital. In addition, three of them, it seems, will fulfil their obligations only after receiving loans from Greece, making their contributions anything but autonomous and raising doubts about the equality of influence which all members of BSEC can expect to enjoy.

Table 3.1 Real GDP

	1989	1990	1991	1992	1993	1994	1995	1996
Albania	9.8	−10.0	−28.0	−7.2	9.6	9.4	8.9	8.2
Armenia	—	—	−12.4	−52.6	−14.1	5.4	6.9	6.6
Azerbaijan	—	—	−0.7	−22.1	−23.1	−18.1	−11.0	1.3
Bulgaria	−0.5	−9.1	−11.7	−7.3	−1.5	1.8	2.6	−9.0
Georgia	—	—	−20.6	−44.8	−25.4	−11.4	2.4	10.5
Moldova	—	—	−17.5	−29.1	−1.2	−31.2	−3.0	−8.0
Romania	−5.8	−5.6	−12.9	−8.8	1.5	3.9	7.1	4.1
Russia	—	—	−5.0	−14.5	−8.7	−12.6	−4.0	−2.8
Ukraine	—	—	−11.9	−17.0	−16.8	−23.0	−12.0	−10.0

Source: IMF, *World Economic Outlook* (Washington, DC: IMF, May 1997), p. 141.

Table 3.2 Inflation

	1989	1990	1991	1992	1993	1994	1995	1996
Albania	—	—	35.8	225.2	85.0	22.6	7.8	12.7
Armenia	—	—	100.3	824.5	3731.8	5273.4	176.7	18.6
Azerbaijan	—	—	105.6	912.6	1129.7	1664.4	411.7	19.8
Bulgaria	6.4	23.9	333.5	82.0	72.8	96.0	62.1	123.0
Georgia	—	—	78.5	887.4	3125.4	15606.5	162.6	40.2
Moldova	—	—	162.0	1276.0	788.5	329.6	30.2	23.5
Romania	—	5.1	161.1	210.4	256.1	136.7	32.3	38.8
Russia	—	—	92.7	1353.0	699.8	302.0	190.1	47.8
Ukraine	—	—	91.2	1209.7	4734.9	891.0	376.0	80.0

Source: IMF, *World Economic Outlook* (Washington, DC: IMF, May 1997), p. 149.

Table 3.3 GNP per capita[1]

Albania	380
Armenia	680
Azerbaijan	500
Bulgaria	1250
Georgia	—
Moldova	870
Romania	1270
Russia	2650
Ukraine	1910

[1] In dollars and for the year 1994

Source: World Bank, *World Development Report 1996. From Plan to Market* (Oxford: Oxford University Press, 1996), pp. 188–9

Crucial delays in the operation of such a key institution are impacting elsewhere, since they are responsible not only for the postponement of other important projects, but also for a more general laxity of the BSEC's activities. At an individual member state level, the Albanian case can give us another dimension of the economic and socio-political complexity of the region. Being one of the most backward countries of the area, its efforts for reform are extremely painful. At the beginning of 1997, the situation ran wild after the collapse of various investment schemes of questionable respectability, the so-called 'pyramids'. A large number of Albanian citizens, attracted by the high interest rates offered, but without having any particular experience in how such schemes work (in order to protect themselves) and following the encouragement of the equally inexperienced (and some say corrupted) Albanian government, invested their savings in these schemes, savings which in many cases were the product of their hard work as immigrants. Unprecedented social unrest burst out when the 'pyramids' collapsed and people came to realize the fraud which had been committed at their expense. With their money gone, the angry citizens turned against the state authorities which were held responsible for this misfortune. Police stations and military camps were attacked, burned and looted of their weapons and supplies, government officials were beaten and persecuted, and soon the Albanian streets (especially in the southern provinces and the capital Tirana) were full of armed gangs in cars and military vehicles (even tanks), shooting in the air and demanding the resignation of the government. This extremely chaotic situation went on for weeks, paralyzing the country – which seemed to have no central authority at all – and led to the intervention of other European countries, which following the invitation of the Albanian government, sent in army units to help restore order. The European countries (among them, Italy and Greece) also helped in the organization of elections, which were eventually won by the opposition party. These traumatic events had repercussions on the obligations of Albania towards the BSEC. Apart from its inability to pay its share to the BSTD Bank and its contribution to the PERMIS, Albania informed the other member states that it could not assume its term of the BSEC Chairmanship, scheduled for the beginning of 1998.[24] As extreme as the Albanian case might be, it is by no means the only large-scale unrest in the former communist bloc (Georgia, for example, has experienced serious internal upheavals), and can show us how difficult and dangerous the transition process can be for the ex-communist states.

Concerning the two countries which do not belong to the ex-communist group – namely Greece and Turkey – problems in the economy are not so severe, but some do still exist. Greece, an EU member state, with a stable democratic political system and experience in the workings of the market economy is the most advanced member of the scheme. However, even it is currently dealing with the economic problems associated with its efforts to participate in the upcoming European Monetary Union (EMU) and in attempting to achieve that target is enforcing strict austerity measures. Turkey on the other hand, experienced an economic crisis in 1994 due, among other reasons, to high inflation (more than 50 per cent yearly), high public-sector borrowing and high external debt. Since then it has been going through a stabilization programme aimed at medium- and long-term structural reforms, like the widening of the tax base, change of the Social Security System and more rapid privatization of state-owned enterprises, amongst other things.[25] Nevertheless, despite the economic problems these two countries may have, their situation is not comparable to that of the rest of the BSEC states. Both these countries can fulfil their financial obligations towards the scheme, and in that respect they represent the most important moving forces behind it (the Russian Federation can also be considered as such).

The economic difficulties, however, pale into insignificance, when one considers the security situation in the region. Three categories of conflicts[26] are still ongoing within the BSEC membership. These are:

- disputes over sovereignty inside the former Soviet Union
- territorial disputes between member states
- conflicts within individual states and between different ethnic groups.

One of the most long-standing and bitter conflicts is that between Turkey and Greece. Their hostility goes far back in history, but even nowadays the tension is constant. Since 1974, the two countries have been close to war at least three times. The first time was in 1974, when Turkey invaded Cyprus; the second was in 1987, when Turkey launched a plan for oil exploration in disputed areas of the Aegean Sea; and the third was in 1996, when Turkey was moved to dispute the Greek rule over a few rocky islands, again in the Aegean Sea. Among the various issues upon which the tension between the two countries is based are, of course: the Cyprus problem; the disagreement over the continental shelf of the Aegean Sea and the operational control of its airspace; Turkey's opposition to the intentions of Greece to expand its territorial waters to 12 miles; the Turkish demand for the demilitarization of certain Greek islands in the Aegean Sea, and more. According to Greece's official defence doctrine, adopted in 1984, Turkey represents the primary threat to its national security.

Other key problems that we must also mention are the conflict between Armenia and Azerbaijan over Nagorno-Karabakh (an area now controlled by Armenia); the fight between the Russian Federation and Chechnya; the tension in Georgia with the Abkhazians and South Ossetians; the tension between the Russian Federation and Turkey over the determination of the routes of the oil pipelines from Baku, the prospective sale of the Russian S-300 missiles to Cyprus, and the obstruction by Turkey of free navigation in the Bosphorus Straits; the tension between the Russian Federation and Bulgaria, over the price and transportation routes of natural gas from the former to the latter, and over their relations with NATO; and the Turkish conflict with the Kurd minority. To all these we can also add the tense relations existing between some member states and other neighbouring countries, such as is the case between Albania and the Yugoslav Federation over Kosovo, and the other between Greece and FYROM, over the latter's name.

This problematic situation concerning the conflicts in the region has been raised many times by high level officials and heads of state or governments of the BSEC member states, but in most cases in a highly diplomatic way. The lack of determination to discuss and confront the security issues in the region, made Georgia's President Shevardnadze emphasize, in the Summit Meeting on 25 October 1996, that these conflicts are a 'monstrous anachronism' and further to stress that 'one is under the impression that these conflicts, that present a formidable threat for the whole region and for the Organization in particular, take place not in our region but elsewhere on the opposite side of the globe ... [and] ... we should bear in mind that conflicts whose number and acuity would not dwindle, are able to blow up the cooperation and integration that have already started'.[27]

Shevardnadze's statement brings us to another problem the BSEC is facing, that of its politicization. Although from the Istanbul 'Summit Declaration' it is clear that the scheme has been created for the promotion of economic cooperation, various member states (like Georgia, Turkey and Ukraine) are in favour of the development within the BSEC scheme of mechanisms designed to prevent or solve security problems. As early as 1992, Shevardnadze had proposed the creation of an 'alliance of Black Sea States', which would include committees comprising foreign and defence ministers and a conflict settlement centre.[28] This idea was unequivocally rejected by the Russian foreign minister. In 1993, a similar idea, for the creation of a collective security system was proposed by Ukraine, again without any success.[29] One main reason why the politicization of the BSEC is being rejected so far, might be the strong feelings of mistrust that exist among some of the member states. Let us take, for example, the Russian Federation and Turkey. In addition to the various tensions mentioned above, another broader geopolitical issue which negatively affects the relations of these two countries, is which one of them will successfully extend its influence to the countries of the region which emerged after the dissolution of the former Soviet Union. It is an open secret among those people who are following developments in the region, that one main reason why Turkey played a leading role in the creation of the BSEC was the extension of its influence to the ex-Soviet Republics of the area, and particularly to those where the Muslim population is the overwhelming majority. On the other hand, Russia which was feeling extremely weak at the time of the BSEC's creation, wanted to participate in the BSEC in order to eventually re-establish itself as the main power in the region. It would have been unwise on Russia's part to allow Turkey to enter these former Soviet territories, under the umbrella of the BSEC security mechanism. Had the security proposal been accepted, Turkey's influence would have been far greater than through economic and trade agreements alone.

Having discussed some of the economic, socio-political and security problems of the BSEC region, it seems safe for us to say that our earlier characterization of the scheme as ambitious, was rather an understatement. A legitimate question though, arises from the discussion so far. Given the problems, why did these countries come together in the BSEC to pursue such a difficult development programme at all, especially when it is obvious that they cannot provide the scheme with the necessary financial means to promote its projects? The answer to this question it seems is particularly related to the ex-communist member states.[30] All the projects which have been included in the BSEC agenda are necessary for the successful transformation of these countries into market economies and for their general development and prosperity. At the same time though, these states do not have the capabilities (neither the financial nor the know-how), to proceed individually to the implementation of such a demanding plan. What they are attempting via BSEC, then, is not different from what other countries around the globe are attempting in the post-Cold War period, namely, collaboration for the common good. In the antagonistic international economic environment of today, even countries with strong infrastructures and an abundance of means, are joining forces in the creation of unions and cooperative organizations in order to cope with the keen competition. Likewise, and following the principle 'unity is strength', the ex-communist states of the Black Sea area came together to meet the challenges of the post-Cold War world. The participating states knew of course that the creation of the BSEC scheme, by itself, was not enough for achieving their targets, because the basic problem (the lack of adequate means) would continue to exist. It was apparent to them from the outset that additional support mechanisms had to be found. That, after all, is why they incorporated in the Istanbul 'Summit Declaration', as one of their main objectives, the aim of wider participation in a 'Europe-wide economic area' and their 'higher integration into the world economy', while the BSEC is described as something which will 'facilitate the processes and structures of European integration'. In practical terms this means that the desire to establish the closest possible relations with European institutions and particularly with the EU, can be considered the main underlying, if only implicit, objective of the Scheme. Furthermore, most of the participating states have expressed many times their wish to become members of the EU. As such a possibility is rather remote for many of the BSEC states, BSEC itself is a useful route by which they can approach the EU institutions and benefit from them, something which individually would again be very difficult to achieve. At every opportunity, the leaders of the member states and the organs of the BSEC are emphasizing how high the EU and European integration is valued by the Scheme.[31] And it is in this context, and for promoting BSEC's interaction with other international economic and financial organizations and institutions, that the heads of state or governments of the participating states agreed in principle at their meeting in Moscow on 25 October 1996, to the strengthening of the legal and institutional basis of the BSEC in order to transform it from a scheme into a regional economic organization, subject to international law. Increased cooperation with the EU and other European and international organizations could have – besides the expected financial and other benefits – additional advantages for the participating states, of no less importance. It '... could create a feeling among the governments of member states of BSEC that they are involved in a process of European integration' while the further development of the BSEC '... could demonstrate to states in Western Europe that states in Eastern Europe are responsibly-minded enough to cooperate and integrate their economies'.[32]

The BSEC Scheme then is an ambitious plan for the promotion of the economic and other forms of cooperation among the countries of the wider Black Sea area. What the longer term prospects of the Scheme are, it is rather early for us to judge though the obstacles are clearly formidable. As to the future, BSEC itself has stated:

> The success of the BSEC process will greatly depend on the political will and the ability of the Participating States to initiate and coordinate viable projects of multilateral economic cooperation, on the degree the BSEC goals will win over the minds of people, as well as on the extent the States will learn the art of working together and finding proper solutions to common problems. It will also depend on the ability to meet future challenges and find points of convergence.[33]

To all of these we must add the need to overcome the feelings of mistrust and bitterness which have been created by the long-standing conflicts in the region, the preservation of which would be an impediment to any attempt at cooperation in the BSEC area.

NOTES

1. F. Sen, 'Black Sea Economic Cooperation: A Supplement to the EC?', *Aussenpolitik*, 44, 3, 1993, p. 281.

60   Subregionalism and World Order

2. Articles 1, 9, *Summit Declaration on Black Sea Economic Cooperation* (Istanbul, 25 June 1992), pp. 3, 4; *The Bosphorus Statement* (Istanbul, 25 June 1992), p. 9.
3. Article 4, *Summit Declaration* ..., 1992, p. 3.
4. Ibid., Article 8, p. 4.
5. Ibid., Article 5, p. 3.
6. *The Bosphorus Statement*, 1992, p. 10.
7. *Summit Declaration* ..., 1992, p. 3.
8. Ibid., p. 6.
9. Ibid., p. 4.
10. The following paragraphs draw heavily upon *BSEC, From Common Interests to Joint Actions* (Istanbul: BSEC, Permanent International Secretariat, 1997), pp. 12–25.
11. Article 14, in *Summit Declaration* ... 1992, p. 5.
12. *BSEC, From Common Interests* ..., *1997*, p. 15.
13. For a more analytical discussion on the environmental problems of the Black Sea, and for the examination of the various collective efforts which have been put forward in confronting these problems, see M.W. Sampson III, 'Black Sea Environmental Cooperation: States and the Most Seriously Degraded Regional Sea', *Bogazici Journal – Review of Social, Economic and Administrative Studies*, 9, 1, 1995, pp. 51–76.
14. O. Ozuye, 'Black Sea Economic Cooperation', *Mediterranean Quarterly*, 3, 3, 1992, p. 52.
15. The discussion on the various organs of the BSEC is based on various official papers, but it is drawn on in particular from, *BSEC, From Common Interests* ..., 1997, pp. 28–36.
16. The EU had financed the work out of the Business Plan with ECU 250 thousand, from the TACIS and PHARE programmes.
17. Article 61 of the Establishment Agreement specifies the conditions for the establishment of the BSTD Bank as the '... deposit of the ratification instruments by at least 6 countries whose initial subscriptions in the aggregate comprise not less than 51 per cent of the initial authorized capital stock of the Bank', in 'Progress Report of the BSTD Bank', *BSEC, Report of the Ninth Meeting of the Ministers of Foreign Affairs* (Istanbul: BSEC Permanent International Secretariat, 1997), p. 127.
18. Observer Status is granted to a state for a period of two years and to international organizations for an unlimited period. After the two years, an interested country can renew its status. With such a status, participation in the BSEC's workings can be open or limited, based on a decision by the MMFA.
19. *BSEC, Meeting of the Heads of State or Government of the Participating States of the Black Sea Economic Cooperation Scheme and the Eighth Meeting of the Ministers of Foreign Affairs of the BSEC Participating States* (Istanbul: BSEC Permanent International Secretariat, 1996), p. 100.
20. See, for example, the 'Plan of Actions for the Implementation of the Provisions of the Moscow Declaration and the Resolutions, Decisions and Recommendations of the Eight MMFA', *BSEC, Report of the Special Meeting of the Ministers of Foreign Affairs with the Participation of the Ministers Responsible for Economic Affairs* (Istanbul: BSEC Permanent International Secretariat, 1997), pp. 91–100.

21. *BSEC, Report of the Ninth Meeting...*, 1997, p. 131.
22. See for example, *BSEC, Meeting of the Heads of States or Government ...*, 1996, pp. 33, 44, 69, 102, 158.
23. For a more detailed discussion on this matter, see R. E. Ericson, 'On Problems of Economic Transition in the Black Sea Region', *Bogazici Journal. Review of Social, Economic and Administrative Studies*, 9, 1, 1995, pp. 23–31; and F. Sen, op. cit., pp. 284–5.
24. For a better appreciation of Albania's economic difficulties, we can mention that according to the budgeted operational expenditures of the PERMIS for the year 1998, Albania's share is just $37 317, while the overall amount needed from the BSEC states has been estimated at $932 925. The operational cost is distributed differently among the member states; Greece, the Russian Federation, Turkey and Ukraine have to contribute $149 268; Bulgaria and Romania, $74 634; and Albania, Armenia, Azerbaijan, Georgia and Moldova, $37 317. For more detailed information on the budget, see *BSEC, Report of the Ninth Meeting ...* 1997, pp. 116–25.
25. OECD, *Economic Survey, 1995–1996, Turkey* (Paris: OECD, 1996), pp. 1–3.
26. F. Sen, op. cit., 283. These categories can often overlap.
27. *BSEC, Meeting of the Heads ...* 1996, p. 70.
28. G.M. Winrow, 'Discussion of Jack Snyder's Article', *Bogazici Journal. Review of Social, Economic and Administrative Studies*, 9, 1, 1995, p. 48.
29. Ibid., p. 48.
30. I make a distinction between the nine ex-communist states, and Greece and Turkey, because the participation of these two countries in the BSEC is dictated by different reasons. Turkey for example, except for the economic benefits it might have from the activities of the BSEC, is very much interested – as I have said – in the possible expansion of its influence to the Muslim countries of the former Soviet Union. Furthermore, its place internationally is very different from that of the nine ex-communist states, being a NATO member and having special relations with the EU. On the other hand, Greece is already an EU and NATO member and in addition to the prospect of the economic benefits that it could gain from the economic activities of the BSEC, is participating in order to prevent Turkey from expanding its influence in the region.
31. See for example, *BSEC, Meeting of the Heads of State of Government ...*, 1996, pp. 111–13; and *BSEC, Report of the Ninth Meeting ...*, 1997, pp. iii, 25, 41, 69.
32. G.M. Winrow, op. cit., p. 49.
33. *BSEC, from Common Interest ...*, 1997, p. 40.

4 West African Subregionalism: the Case of the Economic Community of West African States (ECOWAS)

Stephen Riley

The Economic Community of West African States (ECOWAS), made up of 16 West African states led by Nigeria, is very much a product of a particular time and place. The ideas behind ECOWAS were promoted in the late 1960s, and the subregional project emerged in the early 1970s, when regional and subregional economic integration was thought to be the key strategy to overcome dependence and underdevelopment on the African continent. Self-reliance, African unity, and economic integration were the goals of the early African nationalists and the inheritors of state power in post-colonial societies. These were understandable objectives as many West African states were chronically poor, often land-locked with a poor infrastructure, and dependent upon the export proceeds of a narrow range of primary commodity exports (particularly coffee, cocoa, timber, bauxite, iron ore and other unprocessed minerals). Even after decades of independence, state-directed indebted development, foreign aid and advice from bodies like the World Bank, Sub-Saharan Africa's (SSA) 45 political economies have a total Gross Domestic Product (GDP) less than that of Belgium, and Africa's share of the global total of foreign direct investment (FDI) in the mid-1990s was an average of two per cent.[1] The total GDP of the ECOWAS subregion in the mid-1990s was, at US$63 436mn, the near-equivalent of Ireland's and far less than the GDP of Malaysia (US$85 311mn) or Singapore (US$83 695mn). Table 4.1 illustrates the dimensions of the political economies of the ECOWAS subregion: 16 coastal and hinterland states in West Africa, from Mauritania in the north-west to Nigeria in the south-east.

Table 4.1 ECOWAS

State	1] *Population*: in millions	2] *Surface area*: in 1000s of sq. km.	3] *GDP*: in millions US$ 1995	4] *Real GDP per capita*: US$ 1995
Benin	5.5	113	1,552	1,696
Burkina Faso	10.4	274	2,325	796
Cape Verde	0.4	4	**	1,862
The Gambia	1.1	11	384	939
Ghana	17.1	239	6,315	1,960
Guinea	6.6	246	3,686	1,103
Guinea-Bissau	1.1	36	257	793
Ivory Coast	14.0	322	10,069	1,668
Liberia	2.7	98	**	**
Mali	9.8	1,240	2,431	543
Mauritania	2.3	1,026	1,068	1,593
Niger	9.0	1,267	1,860	787
Nigeria	111.3	924	26,817	1,351
Senegal	8.5	197	4,867	1,596
Sierra Leone	4.2	72	824	643
Togo	4.1	57	981	1,109

Note

Source: columns 1, 2 and 3 are from the World Bank's *World Development Report, 1997* whilst column 4 is from the UNDP's *Human Development Report, 1997*. The total population of ECOWAS is 208.1mn. The total GDP of ECOWAS was US$63 436mn in 1995; by comparison, the 1995 GDP of the United Kingdom was US$1 105 822mn. The economic statistics for Liberia (and, to a lesser extent, Sierra Leone) are unreliable due to the civil conflicts there in the 1990s. ** no figures given.

The creation of ECOWAS also reflected a period when West Africa's most populous state, Nigeria, was growing in both regional African and global influence due to its oil exports, particularly to the North American market. Accordingly, some have suggested that ECOWAS is simply an instrument of Nigerian hegemony in the West African subregion, building upon the new power politics of the post-OPEC oil price hikes of the early 1970s, which gave Nigeria's military rulers both huge personal wealth and political ambitions. Many of the difficulties of the subregional organization in the period since its foundation in 1975 can be traced back to the suspicion that Nigeria was using ECOWAS for its own purposes. In many ways ECOWAS remains effectively tied to the ideas of, and is dependent upon the developing power structures of that historical moment, both of which have been undercut by dramatic social, economic and political changes in the subregion. There is considerable doubt concerning the future viability of ECOWAS. It has become a more market-friendly and security-orientated institution, but it is questionable whether the organization and its most powerful members are capable of responding to both the opportunities of the new regionalism and the new security and economic dangers of globalization.

The aim of this chapter is to consider these issues by examining the development of ECOWAS principally as a hybrid example of Hurrell's notions of regional inter-state cooperation and state-promoted regional economic integration.[2] It is time for a reassessment of ECOWAS given its expected central role as one of the four pillars of an African Economic Community in the next century and the assumption that it will be the sole subregional organization.

ECOWAS involves often uneasy and partial cooperation between the 16 member states, but its early development was much influenced by the European model of regional economic integration. The chapter proceeds to outline the origins of ECOWAS and the challenges to the development of the subregional project posed by competing subregional and external power blocs and organizations, as well as diverse cultural and ideological forces. Beset by a variety of funding, staffing, political and other impediments, ECOWAS has made little progress on its economic integration agenda. Intra-ECOWAS exports have grown from 3 per cent in the mid-1970s to only 6 per cent in the mid-1990s.

The subregional organization has also changed from being driven by economic goals to being a security-driven institution in the 1990s with the direct military involvement of ECOWAS in the conflicts in Liberia and Sierra Leone. Much ECOWAS activity in the 1990s has been political and security-orientated, even though the revised ECOWAS treaty envisages its new role as the 'sole economic community in the region' and one of four foundations upon which an African Economic Community (AEC) is to be built in the next century.[3] Unofficial economic, social and political cross-border penetration, with the large trade in illicit drugs, guns, workers, and scarce commodities, is more effective than state-led economic integration. Ever greater Nigerian influence over economic and security issues, rather than real cooperation, appears to be at the heart of the organization.

The West African state system – the Westphalian pillars upon which such subregionalist schemes are erected – is in serious decay. The case of ECOWAS thus illustrates the view that there are considerable and diverse obstacles to creating an effective, functioning subregional project in Africa. The most likely scenario is that despite such ambitious subregional projects, an increasingly fragmented continent will undergo endless reconfigurations of state power – with all their bloody consequences – as the rest of the world looks the other way.

THE ORIGINS OF ECOWAS

General Yukubu Gowon, Nigeria's military ruler in the early 1970s, is usually credited with being the creator of ECOWAS, along with the helpful backing of President Gnassingbe Eyadema of Togo. Both army strongmen announced their support for a prospective economic community in West Africa after a state visit by Gowon to Togo in April 1972. Their establishment of a group of experts to examine possible areas of cooperation, and their subsequent lobbying of the respective Anglophone and Francophone blocks of state elites in West Africa, eventually led to the signing of the ECOWAS Treaty in Lagos (Nigeria's then capital), in May 1975 by 15 West African Heads of State.[4] The political leaders represented at the signing ceremony came from a diverse group of countries, only newly independent from British, French, and Portuguese colonial rule. They showed a surface commitment to differing beliefs and doctrines, ranging from the pro-Western, export-orientated developmental capitalism of some of the older Francophone leaders, such as Felix Houphouet-Boigny of the Ivory Coast and Leopold Senghor of Senegal, to the African nationalist Marxism of the former Portuguese colony of Guinea-Bissau. Military junta figureheads, such as Gowon, mixed with democratically elected heads of state, including President Dauda Jawara of The Gambia; African Marxists mixed with Muslims. In addition to diverging ideological viewpoints, the Heads of State presided over regimes with economies and cultures of varying sizes and strengths and links with the former metropoles. The currencies of the former Francophone states, apart from that of Guinea, were linked with the French Franc through the institution of a common monetary zone (the Communauté Financière Africaine – CFA – zone). Looser relationships typified Anglophone West Africa's relations with the United Kingdom, whereas former Lusophone West Africa (Cape Verde and Guinea-Bissau) had just severed links with Portugal after a prolonged and bloody war of national liberation, principally in the interior of Guinea-Bissau.

The 1975 Treaty came into operation when a series of protocols implementing it were subsequently signed in Lome, Togo's capital, in

November 1976. ECOWAS as an institution started in March 1977, and the sixteenth member of the organization, the Lusophone, exceptionally poor and volcanic Cape Verde islands, joined that year. Although the self-effacing Gowon is often credited with originating ECOWAS, ideas of economic cooperation in West Africa, and Africa in general, have a much longer history. Some African nationalist historians evoked early pre-colonial trade across West Africa as a model to emulate post-independence. There were informal discussions about greater cooperation amongst the newly independent states from the mid-1960s onwards, often involving the idea of free trade between members, as proposed by President William Tubman of Liberia in 1964. These ideas and similar initiatives by the United Nations Economic Commission for Africa (ECA) echoed the dreams of some of the earliest PanAfricanists, such as the veteran Ghanaian nationalist Kwame Nkrumah, who was an advocate of both the merger of the artificial post-colonial states into larger more economically viable entities, and a proponent of continental unity.[5]

The subregional institution that was established had elements of both regional inter-state cooperation and state-promoted regional economic integration. It was headed by the 'Authority of Heads of State and Government' (usually called the ECOWAS Summit) which meets once a year and a subsidiary Council of Ministers which meets twice a year. The Summit has proved useful for West African leaders: it is a media event and usually receives fairly wide coverage in the international press. It also enables small groups of leaders to meet together privately to resolve disputes, such as the meeting between the leaders of Ghana and Togo at the 1988 Summit which settled a dispute over their mutual expulsion of nationals.[6] As ECOWAS has established itself, it has developed a series of formal and *ad hoc* commissions, groups and committees to deal with specialized functions and particular crises. Commissions have been created to deal with trade, monetary arrangements, transport, resources, social and cultural affairs, and information. Newer elements of ECOWAS include the ECOWAS Monitoring Group (ECOMOG), which oversaw the involvement of ECOWAS in Liberia's civil war from 1990–97; and the ECOWAS Committee of Five, which, following a coup d'état and rebel takeover of state power in May 1997, is seeking to mediate in the conflict in order to restore the democratically elected government of Sierra Leone.[7] This outline of the institutions of ECOWAS illustrates its trajectory: it was initially concerned with the liberalization of trade as the means to secure economic integration but has become increasingly concerned with security issues. Specific provisions of the Treaty of Lagos called – unrealistically – for the establishment of a free-trade area by 1990 and for the subsequent creation of the conditions for the free movement of people, services and capital within ECOWAS. Neither was achieved. Another example of slow movement was the original intention to create a single monetary zone by 1994. As this had made no progress by July 1992, the target was changed to the year 2000. The 1997 Summit established an *ad hoc* monitoring committee to oversee moves towards this goal. It would involve a key role for the ECOWAS Fund for Cooperation, Compensation and Development (the ECOWAS Fund, based in Lome, the capital of Togo), which has a degree of autonomy and finances projects which help integrate the subregion, as well as compensate for lost customs revenue. Anthony Ani, Nigeria's Finance Minister, argued at the 1997 Summit that his country's currency, the Naira, could become the subregional means of exchange.[8] This is extremely unlikely, as few of the Francophone states would wish to give up the monetary stability provided by the link between their common currency and the French Franc, unless France changes the fundamentals of that relationship.

THE ECOWAS INSTITUTIONS

ECOWAS has an Executive Secretariat which was established in Lagos and subsequently moved to Abuja, the new Federal capital of Nigeria. As with many African international organizations, the ECOWAS Secretariat has faced a series of problems: some personal, factional or political; others financial. The Secretariat has had to grapple with personnel difficulties due to patronage appointments dictated by member governments, and squabbles between some of the principal officers. It has also lacked independence from its members, particularly the most powerful one, Nigeria. What is more, the Francophone–Anglophone rivalry in West Africa has undermined the coherence and efficiency of the organization. One particularly lacklustre Executive Secretary was replaced halfway through his eight-year term, partly as a consequence of the criticism that the ECOWAS Secretariat was failing to implement its mandate. But the Secretariat has had major problems trying to deliver on its mandate, including financial ones. By 1992 the arrears on payments owed by member governments amounted to US$28.0mn on an annual expenditure of US$10.0mn. At the 1997 summit, the arrears had reached US$44.0mn and the organization was planning a community levy (on the total value of imports) to replace the unpaid direct budgetary contributions by member governments. The only fully paid-up members were Benin, Ivory Coast and Nigeria.[9]

In addition to political pressures upon ECOWAS by member

governments, the organization has faced problems of institutional capacity-building, due to the separation of elements of the Secretariat and the ECOWAS Fund, and continuing difficulties with a bewildering number of other regional and subregional international organizations, which often share overlapping responsibilities. Just prior to the creation of ECOWAS, the Francophone West African states, with the help of France, transformed a moribund regional organization into the Communauté Economique de l'Afrique de l'Ouest (CEAO). The CEAO's purpose was to create a single customs union, free up the movement of goods and capital, and encourage cooperation for industrial production in Francophone Africa. Georges Pompidou, the French President at the time, said that it would produce a 'just equilibrium' between the Anglophone and Francophone states in the subregion.[10] It also reinforced the traditional rivalries in the subregion: particularly between the more powerful Francophone states, such as Senegal and Ivory Coast, and the influential Anglophone states of Ghana and Nigeria. However, these manoeuvres prior to its foundation undermined one of the purposes of ECOWAS, which was to try to bridge the economic, social and political distances between the former colonial possessions.

Similarly, the negotiations to revise the ECOWAS Treaty in 1992–93 were accompanied by the expectation of changes in the international organizations of the Francophone block in West Africa. It was initially thought that the West African Francophone states would be willing to merge their organizations into a revitalized ECOWAS. However, the revival of repressive militarism in Nigeria in 1993 with an aborted democratization process, and the difficulties of the CFA common currency zone in 1994, after a 50 per cent devaluation, created obstacles to such a merger. Francophone states subsequently established their own new union, the Economic and Monetary Union of West African States (UEMOA) which is being externally guaranteed by France. This development upset the plans of ECOWAS to move towards a single economic community and gave aid donors an alternative route to channel aid to the subregion. Many economists argue that the key to a successful economic community or an attempt at regional integration is the establishment of strong links with an external guarantor, but this contrasts with aspirations towards African self-reliance, and creates anxiety about, or hostility towards, such forms of external dependence.[11] Whatever the theoretical argument, ECOWAS was diminished by this rivalry.

The revised ECOWAS Treaty was adopted by the Summit of Heads of State in July 1993, although there were delays in ratifying it. The Treaty reflected both the mood of the moment and also illustrated the weaknesses of the original Lagos Treaty. The governing principles of ECOWAS were revised and updated and they were in tune with the fashionable international donor policy consensus of the time on issues such as the environment, markets and the private sector, regional integration and good governance. The Treaty's guiding articles echoed the movement of the European Union (EU) towards economic integration, although the revised Treaty did not accept the 'variable geometry' principle of the EU, and expected a coherent and concerted movement towards a regional African Economic Community by 2025 (as envisaged by the OAU's Treaty of Abuja in 1991). Controversially, it also added a political and ideological dimension, emphasizing the value (at least in the abstract) of democracy, administrative efficiency and conflict resolution. The two defence protocols were to be reviewed, reflecting the preoccupation of West African states with the new security issues. Article 3 of the new Treaty stressed the peaceful settlement of disputes, the value of human rights, and the principles of political accountability, economic justice and participation.[12]

DOCTRINE AND IDEOLOGY IN ECOWAS

The ideas behind the development of ECOWAS are fairly orthodox. The Lagos Treaty reflected the European experience of the time with its emphasis upon free trade within a common external tariff area. Another key idea was the free movement of workers, services and capital inside ECOWAS. The pursuit of economic integration within the subregion was expected to have damaging consequences for some states and thus compensation for lost customs duties, as well as equalization in development, was to be achieved through the activities of the ECOWAS Fund. The 1993 revised Treaty sought to move ahead on economic integration and rationalize the thirty subregional inter-governmental organizations with responsibilities in this area. For obvious reasons, the rhetoric of ECOWAS was often inconsistent: appeals to African sentiments and expressions of suspicion about international markets and the 'West', fitted badly with a rationale which was tailored to generate further supplies of external aid and credit.

Whatever the rhetoric, the ECOWAS institutions faced an overwhelming task in one of the world's poorest and most socially divided subregions. Eight of the fifteen lowest ranking states on the UNDP's Human Development Index are in ECOWAS. But the respective state elites look not to their own citizens but to Paris, Lisbon or London (and, increasingly, to Washington, DC): their tastes, ideas on development, their cultural references, and their politics, were and are influenced by the metropole, although colonial rule ended in most cases in the late 1950s or early 1960s. However, beneath the governing cliques, oppositional cultures began to emerge which sought to articulate an alternative to the state-led trade liberalization of ECOWAS and were increasingly hostile to the neo-patrimonial politics of the central state in West Africa. Informal trade, survival strategies based upon opting out of conventional economies and the harsher effects of structural adjustment, and the politics of a growing and alienated youth all combined with the uneven effects of globalization. Oppositional cultures developed in the 1980s and 1990s in major cities such as Dakar in Senegal and Lagos in Nigeria. In some states, these oppositional cultures had a direct political dimension and underpinned political action, as with the case of the youthful members of the rebel forces (so-called 'child soldiers') in Liberia and Sierra Leone. In other cases, they led to the development of what Bayart calls the 'culture of derision' drawing upon West African and foreign music, literature and other cultural forms.[13] Neither tendency supported the ECOWAS project.

Several ECOWAS members remained more committed to either the former metropole or to a state-led model of development. In the 1990s, at the level of doctrine, ECOWAS itself became more sceptical about the state-led model and increasingly emphasized the private sector as the vehicle for the achievement of its goals. In so doing, they reflected the policy nostrums of the principal Western aid donors, including the World Bank, with its emphasis upon a reduced role for the state in the context of future externally orientated development. The practice of ECOWAS, however, was rather different as the subregional project was in essence based upon the state system. In any case, as Leys has argued, and as many West African political leaderships realized, 'the African state, for all its record of abuse, remains a potential line of defence for Africans against the depredations of the world economic and political system.'[14]

ECONOMIES: OFFICIAL AND UNOFFICIAL

ECOWAS has had little effect upon trade, investment and production in the subregion: most economic activity is unaffected by the organization and its goals. Instead, trade and other economic patterns reflect well-established colonial and pre-colonial links. Although ECOWAS has been operating since the mid-1970s, there has for example been very little change in trading patterns. Of the various tests of regional integration that have been suggested, perhaps that of de Mello and Panagariya is the most useful.[15] Arguing that regional integration schemes should be discouraged, they suggest that, for a regional integration scheme to be viable, intraregional exports should be above 4 per cent. ECOWAS passes this test in the 1990s, with intra-ECOWAS exports at 6 per cent, although this is due in part to substantial and unavoidable trade between the coastal states and the hinterland states (such as Nigeria and Niger, or Ivory Coast and Burkina Faso or Mali).

Both the well-being of West Africa's people and state revenues are based upon long-established patterns of externally orientated trade. This has included crude oil exports from Nigeria, the trade in unprocessed coffee and cocoa from Ghana and the Ivory Coast, and a wide variety of other raw minerals or materials exports: alluvial gem diamonds, rutile (or titanium dioxide) and gold from Sierra Leone, iron ore, timber and rubber from Liberia, bauxite from Guinea, and uranium ore from Niger. These export trades have fluctuated due to movements in world market prices, and both economic and political instability in the exporting states in West Africa. Virtually all states, with the partial exceptions of Nigeria and Ivory Coast, have very low levels of industrialization, although ECOWAS has tried to encourage such developments, amidst a wide variety of other projects, including those in transportation, agriculture, and drought and desertification. A review of ECOWAS initiatives in the early 1990s found that only 31 of 136 had secured funding and there were immense difficulties in coordination between the ECOWAS Secretariat and national governments. There were few notable successes. Nevertheless, some schemes, such as the West African coastal road network, and the trans-Sahelian highway, were virtually complete.[16] Ironically, the schemes that were most successful were those least attuned to the integrationist goals of ECOWAS.

A major continuing problem for coordination is the economic as well as political and social heterogeneity that has characterized the states of the subregion since independence in the late 1950s and early 1960s. Several Afro-Marxist regimes, such as Luiz Cabral's Guinea-Bissau, 1974–80, and possibly Sekou Toure's Guinea, 1958–84, have been replaced by regimes of a more conservative hue and now espouse market liberalism and pluralist politics. Although there were radical populist regimes, such as Thomas Sankara's Burkina Faso, 1983–87, and later synthetic versions in The Gambia (post-1994) and Sierra Leone (1992–96), they contrast markedly with the conservative Muslim politics to be found in the (semi-detached) Mauritania, and Senegal. Genuine civil pluralist politics persisted in The Gambia from 1965 until 1994, in Senegal since the mid-1980s, and briefly in Nigeria, during 1979–83. The more recent democratization of some African states in the early 1990s has coexisted with varieties of military authoritarianism in most West African regimes since the 1970s.[17] Whatever the political form, the economic costs of politics have been high with many instances of swollen and inefficient states and gross examples of waste, mismanagement and corruption.

Many West African states have harboured ambitious development plans that have been thwarted by a variety of internal and external forces. Outwardly orientated export projects favoured by the World Bank have more recently replaced often grandiose import substitution industrialization schemes. But all regimes have had to come to terms with the politics and economics of economic globalization. There was an expectation from the early 1980s onwards that the regimes would reform both economically and politically as aid budgets shrank, external debts grew and the terms of trade in their primary commodity exports deteriorated. A consequence of this was the impact of externally imposed structural adjustment programmes upon state strategies and such subregionalist projects. Structural adjustment has had profoundly damaging consequences for the state-led subregionalism envisaged by the founders of ECOWAS. It undermined the centrality of the state to the development project and the imposition of adjustment in one society had destabilizing effects in its neighbours. As Daddieh has pointed out, structural adjustment programmes are, 'incompatible with the objective of promoting collective self-reliance through regionalism because they have a tendency to reinforce ... Africa's historical role in the international division of labour and the extroversion of its political economies'.[18] In fact what has happened is that the principles of the subregional project have changed to reflect the priorities of structural adjustment which requires a reduced state role in the economy, and an outward-orientated, free-market development policy.[19]

The contemporary economic dimensions of subregionalism in West Africa have been affected by two developments which have stymied the project: the growth of informal or parallel trade, and the continued French presence in politics, trade and monetary arrangements. Paradoxically, although there has been virtually no growth in official trade, there has been a huge growth in parallel, smuggled or informal trade in the subregion. Well-established informal trade patterns have been supplemented by newer trades in primary commodities or consumer items where a preferential advantage is to be gained by moving the goods across a national border, often a consequence of structural adjustment or inefficient state regulation. The Gambia, Ghana, Nigeria and Sierra Leone are all centres of systems of extensive informal trade. The small riverine state of The Gambia is surrounded on three sides by Francophone Senegal and is the base of a huge smuggling economy involving several states. A strong Gambian currency and very liberal import policies undercut the otherwise strong CFA franc: consumer imports were exchanged for Senegalese groundnuts, for example, in the 1970s and early 1980s. A later liberalization of the Gambian economy from 1985 onwards meant huge increases in the import of rice into The Gambia for the onward illicit re-export into Guinea-Bissau, Mali and Senegal, but the CFA devaluation of early 1994 has also altered the dynamics of this new illicit trade system.[20]

Ghana was once at the centre of another large parallel economy which linked an Anglophone with several Francophone countries, due to the overvaluation of the Ghanaian cedi. Mainly cocoa, but also unprocessed minerals such as gold, were unofficially exported into Togo and Ivory Coast in exchange for cigarettes, alcohol, vehicles and general consumer goods, especially in the mid- and late-1970s as Ghana's economy deteriorated.[21] The liberalization of the Ghanaian economy from 1983 onwards has not succeeded in eradicating this trade. Diamonds continued to be smuggled out of Ghana, and a free-flow parallel trade continues on the borders with Togo and Ivory Coast.

A protectionist Nigeria in the 1980s and 1990s created an informal market where consumer goods, such as cigarettes, alcohol and electronic equipment, flow into the country, and subsidized petrol, fertilizer and agricultural produce, including grains and cocoa, flows out. The outflow secures funding in the harder CFA franc currency to finance smuggled imports. Customs statistics from the neighbouring states suggest that the illicit trade over Nigeria's borders involves billions of CFA francs, even after the liberalization and devaluation of the Naira after 1986.[22]

In all these and other West African cases, there is a widespread suspicion and some evidence that prominent politicians, civil servants, and their commercial allies are actively involved in the trades. For example, in Sierra Leone, the growth in the parallel economy was linked to the systemic corruption associated with the personalist rule of President Siaka Stevens (1968–85) during the initially one-party dominant and later one-party All People's Congress (APC) regime of 1968–92. It rested upon the theft of government revenues and the unusual characteristics of the political economy of Sierra Leone, an economy principally based upon the official, and in the late 1980s and 1990s mostly smuggled, export trade in alluvial gem diamonds and gold, and agricultural produce, including coffee, cocoa, and palm products. From the late 1970s onwards, a huge parallel economy developed, which exceeded the official economy in size, and involved political middlemen and Lebanese and Indian traders. These developments were later conceptualized as a 'shadow state' where a pattern of corrupt rule existed behind the formal façade of political power, based upon informal markets.[23]

The late 1980s and 1990s has seen the emergence of new elements of these illicit markets. New trades in illicit drugs, including heroin and cocaine, and light weapons and the other merchandise of war, have joined the older established trades. The corrupt income for the public officials involved is substantial. This growth in corrupt income in West Africa is associated with the continuing problem of the external debt of the states and global capital flows. In the early 1990s, Nigeria became a major transhipment centre in the international illicit trade in cocaine and heroin in addition to its continuing problems with high-level public sector corruption. The proceeds of these trades across West Africa as well as corrupt income in general are siphoned off into offshore tax havens and conspicuous consumption.[24] Whatever the size and end destination of these illicit trades, and the corruption that is often associated with them, they do not contribute to the ECOWAS project of state-led subregional economic integration.

The French cultural, political, economic and military presence in West Africa has also had long-term and damaging effects on the ECOWAS project. France's links with the region are tied in with French elite perceptions of their country's role in the world and are as a result long-standing and intense. One recent study has argued that, 'although the tricolour came down over Africa, France never really relinquished control of its possessions'.[25] France has maintained its long-standing ties operating through pro-French, conservative political leaders (such as Houphouet-Boigny in Ivory Coast), contingents of French paratroops stationed permanently in the major West African Francophone states, economic interlinkages, and close political ties between the Francophone elites and the French Presidency orchestrated by a system of fixers and secret service agents headed by Jacques Foccart, a close aide to several French presidents.[26] One reason for the attractiveness of a 'neo-colonial' relationship between the Francophone West African community and France is the economic stability that such an arrangement brings: France may gain secure supplies of uranium ore for its nuclear power-dependent electricity system, and markets for its manufactured goods, but the Francophones in West Africa also gained. They gained aid and technical assistance, recognition, economic stability because of the CFA franc zone, and secure access to western markets. This relationship, and the relationships between the other former colonies and their metropoles, varied considerably from state to state. Central to the relationship are the four core Francophone states of Cameroon, Gabon, Ivory Coast and Senegal.[27]

The mid to late-1990s have seen a gradual reduction in the French influence in West Africa, as France has finally come to terms with its overextended post-Imperial ambitions along with a recognition of the escalating costs of membership of the European Union and the European Monetary Union [EMU]. French commerce, and the French military, have started to retreat from West Africa. Despite the economic gains of the devaluation of the CFA franc in 1994, and a French trade surplus of US$1.0bn a year in the mid-1990s, French commerce seems reluctant to invest, unless there is a substantial and continued upturn in West African economies. The French military aid budget to West Africa has been reduced by 30 per cent since 1992, and France is now more sympathetic to ideas of an all-African security force to replace French security agreements with a number of Francophone states by 1999–2000.[28] Symptomatic of the end of this post-Imperial overstretch was France's brief and counter-productive central African intervention in Rwanda in 1994, and its continued links with Zaire until the collapse of Mobutu's regime in 1997. These embarrassments were a prelude to the new Socialist government's announcement of a rethink of Francophone policy in mid-1997. The retreat of France, especially from its major former colonial possessions of Senegal and Ivory Coast, thus gives Nigeria the opportunity to play a much stronger role in West Africa.

THE POLITICS OF SUBREGIONALISM

One of the key features of ECOWAS in terms of world order is the significance of Nigeria as a hydrocarbon state: it owes its predominant position in the subregion to its extensive oil reserves, which gives it economic and political power and underpins its growing security role in the 1990s. Nigeria is a classic rentier state, with most of its wealth since the early 1970s being a product of the taxation of exploration for, and exploitation of, crude oil. In Nigeria, the key political relationships are thus between the power-holders (principally senior military officers, for all but four years of the existence of ECOWAS) and the extractive industries (with a central role held by oil MNCs such as Shell). Nigeria's more than one hundred and ten million citizens represent at least a third of total ECOWAS GDP. But for many years, the key Nigerian issue has been the politics of competing military power-blocs. The Nigerian citizenry has been systematically excluded from power. The state's revenue is over-reliant upon oil proceeds. The military (and other short-lived civilian) governments have curried favour and enriched themselves by distributing the largesse from oil extraction, rather than by broadening representation in Nigerian politics. Instead of citizenship politics and representation, the key political issue has been unequal redistribution in the context of a highly corrupt spoils system.[29] Thus a defining characteristic of Nigerian politics has been redistribution: away from the oil-producing areas (such as the lands of the Ogoni in the south-east) and to the more politically powerful, and Muslim, north.

However, there have been ebbs and flows in Nigeria's hydrocarbon power, dependent upon broader currents in the global political economy. Nigeria's current external indebtedness, at a record US$34.29bn, is a product of a continuing stagnation in world oil prices from the high-water mark of the OPEC activism of the early and late 1970s. But Nigeria does remain by far the most important subregional actor: it has both military might, with the largest standing army in the subregion, and an overpowering broader economic influence. Nigeria may be heavily indebted externally but the military junta of Ibrahim Babangida and his successor, Sani Abacha, in the 1990s, have still been able to fund the ECOMOG intervention in Liberia and interventionism elsewhere.[30]

For West Africa in the mid- and late-1990s, and into the next century, the central issue is thus likely to be the continuing significance of this 'Pax Nigeriana': Nigeria as a subregional security power, and its regionalist ambitions in Africa. Abacha's regime used the subregion to achieve its domestic and international goals. In terms of domestic politics, Abacha's external military interventions (for example, in Liberia, under the auspices of ECOMOG) meant that potentially rebellious troops, and potential coup plotters, were kept occupied outside Nigeria. They also benefited from the informal economic opportunities that military interventionism brought. Nigerian members of ECOMOG were accused of widespread looting in Liberia. Critics suggested that the acronym ECOMOG stood for: 'Every Car Or Moving Object Gone'.

The junta also had a series of broader regional and international goals: Nigeria is seeking to make gains in the wider African region and is seeking to offset criticism of its human rights record by the West. Whilst the Nigerian junta gained some diplomatic advantages, it was unable to deflect continuing human rights criticism on a range of issues, including the execution of Ken Saro-Wiwa and other Ogoni environmental activists in 1995, and the continued detention of former military officers and politicians such as Olesugun Obasanjo (ex-head of state), and the late Moshood Abiola (winner of the unrecognized 1993 Presidential elections). Despite these issues, there is very little that external powers can do to restrain Nigeria's growing – and often malign – influence on the subregion.

Nigeria's influence upon the ECOWAS subregional project is also affected by a range of external influences. These include the relative strengths of the other significant powers in the subregion, particularly Ghana, Ivory Coast and Senegal. The other states, such as the very small Lusophone countries, count for very little. Nigeria's relations with the other West African states has often been fraught. There has been continuing friction with various regimes in Ghana, Togo and Niger over migrant labour. One of the early decisions of ECOWAS was to try and free up the labour market by allowing for the free movement of individuals. Even though Article 27 of the Treaty of Lagos expects the abolition of obstacles to freedom of movement within the Community, and a subsequent 'Protocol on Free Movement of Persons' was signed and ratified, there have been continuing difficulties in this area. Community citizens, supposedly exempt from visa requirements, have been denied entry to other states. Nigeria has expelled huge numbers of illegal aliens, principally Ghanaian nationals, twice: during 1983 and 1984. The 1983 expulsion affected more than two million people, who were forcibly and rapidly expelled causing great distress and chaos in the neighbouring states. Other states, such as The Gambia, Ghana, Guinea and Sierra Leone, have also expelled nationals of other Community states.[31]

SECURITY AND INSECURITY IN THE SUBREGION

> We are entering a bifurcated world.
> Part of the globe is inhabited by Hegel's and Fukuyama's Last Man,
> healthy, well-fed and pampered by technology.
> The other, larger part is inhabited by Hobbes's First Man,
> condemned to a life that is 'poor, nasty, brutish and short'.
> Although both parts will be threatened by environmental stress,
> the Last Man will be able to master it; the First Man will not.[32]

It was Robert Kaplan's brief visits to the West African states of Guinea, Liberia and Sierra Leone – the places where the First Man (and Woman) dwell – that gave rise to his ideas on the new forms of insecurity in the post-Cold War era. He argued in 1994 that a Hobbesian future was likely for most of the globe, akin to what is occurring in West Africa at present. The crude, low-technology violence, seen in many of West Africa's conflicts, was, according to Kaplan, lacking a political motive. Instead, the motivations of the atomized and socially alienated West African youths involved was simply primitive accumulation, or less euphemistically, crime or theft. Environmental stress, rapid population growth and mass migration will lead to the collapse of communities and nation states. These views were shared by other writers.[33] Conflating the ideas of Van Creyveld and Homer-Dixon, and applying them to West Africa, Kaplan argued that internal wars will increasingly be intra-state in character and fought over resource issues under the guise of religious or ethnic conflicts.

Many have criticized Kaplan's views, particularly the environmental and demographic determinism involved and the shallowness of his understanding of West Africa.[34] But there are new security threats emerging which challenge West Africa's system of fragile and artificial post-colonial states, with their often weakly authoritarian or quasi-democratic leaderships – a state system which is the principal means by which the ambitious goals of ECOWAS can be delivered. Youth politics, the easy availability of light weapons, the temptations of crime, the profits from both large-scale and small-scale illicit drug and other commodity smuggling, urbanwards migration, and the desperation which is a product of the social damage that economic decline and structural adjustment cause, have all contributed to a deterioration in security in the subregion in the 1990s. Nigeria has become one of the global centres of the illicit drug trade in the 1990s and there are escalating problems of both crimes by the powerful (such as massive frauds and transnational corruption) and violent crime against the powerful. Most wealthy people have domestic security and pay off the local police. Environmental and the other 'new' issues in the redefined security agenda of the 1990s are thus of great importance to West Africa as a region. There are few signs that ECOWAS has the political will or the ability to deal with these new problems.

As well as the new security threats, there are a good number of more conventional ones. Many West African states depart from Weber's overused but still significant definition of the state as the monopolistic and legitimate wielder of the means of violence. Over-large, socially divided and often rebellious conventional armies (as in Nigeria), foreign forces (such as French Paratroops, and mercenary forces including the South African 'Executive Outcomes'), paramilitary or 'traditional' militia forces (such as the Kamajor militia in Sierra Leone or elements of the Tuareg in Mali), armed secessionist groups (such as in Senegal's Casamance region), and the armed elements of 'Warlord politics' (as in Liberia) are factors in the new security. Although top-heavy conventional armies consume a good deal of the state's resources, despite the absence of any significant external threats, they are often relatively weak when trying to maintain order and internal security against this range of challenges to central state power. In

Liberia in 1990, a regime with a supposedly strong army was toppled within nine months of a challenge being launched. Often the security situation in 'complex emergencies', such as those in Liberia and Sierra Leone in the period 1990–97, involves several competing threats to the conventional army: foreign troops (from Nigeria, Ghana, Guinea and The Gambia), foreign security forces (including the Gurkhas), armed militias and 'Sobels' (in Sierra Leone: 'soldiers by day, rebels by night').[35]

ECOWAS has been concerned with the more conventional security issues since its foundation, although the Lagos Treaty did not have any specific reference to cooperation on defence or specify that the organization would have a peacekeeping role.[36] As with many of the developments in ECOWAS, there were later negotiations on common defence and security amongst the Heads of States at subsequent ECOWAS Summits. A defence pact was eventually agreed after the 1980 Summit although several states dissented. It involved the expected creation of a Defence Council made up of the ministers of defence and foreign affairs of member states, and attempts to constitute an allied force of the Community, although there is no standing army.[37] Early attempts to use the defence pact, for example in relation to border skirmishes between Mali and Burkina Faso in 1985 (over disputed resource-rich territory), and Senegal and Mauritania in 1989 (over mutual expulsions of nationals), proved ineffective.

The ECOWAS intervention in Liberia from 1990 onwards thus marked a radical departure. It is one of the few instances where a regional peacekeeping force has been sent to preside over both the arbitration between warring factions and a successful process of conflict resolution, which eventually led to the end of the civil war with a brokered peace settlement and an election in mid-1997. General Sani Abacha claimed that Nigeria's role in the conflict, at the core of the ECOWAS mission, had cost the lives of many Nigerian soldiers and more than US$3.0bn.[38]

The ECOWAS intervention in Liberia took place on the basis of the 1981 defence pact, although key elements of that pact (such as the establishment of a Defence Council) were not in place, and the pact itself was not ratified until 1986. Initially, the Francophone members of ECOWAS were reluctant to interfere in the conflict, which started when a small rebel group headed by Charles Taylor, an exiled Liberian dissident, entered Liberian territory at the end of 1989 and challenged the government of Samuel Doe. It took less than nine months for Taylor to topple Doe's regime and the country then split into warring factions. After a bloody civil war which has destroyed Liberia's development prospects for many years, Taylor was elected President of Liberia in the 1997 elections.

ECOWAS sought to mediate in Liberia in mid-1990 with the creation of a 5-member mediation committee which first called for a cease-fire and then established the ECOWAS Monitoring Group (ECOMOG). The purpose of ECOMOG was to separate the warring factions and keep the peace. Nigeria's military leader, Ibrahim Babangida, committed Nigerian troops; they were joined by troops from three other Anglophone countries (Ghana, Sierra Leone and The Gambia). Nigeria has provided most of the troops and the funding for ECOMOG for all the seven years of its operation.

The subsequent Liberian conflict has seen cyclical periods of relative peace and negotiations, and renewed conflict between the 'Warlords' (such as Charles Taylor) and ECOMOG.[39] ECOMOG evolved from an independent mediating force into an active participant in the conflict. Nigeria's military contingent grew from 5000 (out of 10 000) troops to 9000 (out of 12 000). In addition to Nigeria, other West African states also played a mediating role, especially as the key Liberian political figure, Charles Taylor, was adept at exploiting the hostilities between the Anglophone and Francophone members of ECOWAS. The Francophone states of Guinea and Senegal also sent troops (with American financial support) in the early years of the conflict.[40] The Ivory Coast, an initial supporter of Taylor, was a mediator from 1993 onwards, and Ghana sought to strike a deal between Taylor and the Nigerians. The task of ECOMOG became increasingly political rather than military, particularly when an agreement was reached after Taylor's troops fought for control of the capital, Monrovia, in April 1996.[41]

The outcome of the conflict was a victory for Charles Taylor's faction, which was legitimized by the 1997 election. It is a successful if strange example of subregional cooperation. Although there were human rights abuses and corrupt accumulation by the ECOMOG force, the intervention was a bold effort to resolve a local conflict locally, with minimal external involvement. A customs union had become a peacekeeper and peacemaker, although there are few signs that this success can be repeated elsewhere in the subregion. ECOMOG was highly dependent upon Nigerian political commitment and financial backing. Such a subregional initiative is unlikely in the future unless Nigeria's interests are served by it.

CONCLUSION

West Africa is of marginal interest to the core western capitalist states. In terms of world trade and geopolitics, ECOWAS is a group of small, primary commodity exporters with little political influence or strategic significance, a relatively small market for western exports, and virtually no possibility of emerging as an industrial export challenge to the old or the new 'late, late' industrializers with a developmental state providing sustained economic growth. The over-ambitious subregional plans for the economic integration of ECOWAS are of little consequence. Only Nigeria's crude oil exports are of significance to the United States, and Niger's uranium to France, although France's export trade with the subregion remains of some importance in the late 1990s. From the West's point of view, the subregion instead represents an arena of social and political instability, exceptionally weak states, and growing economic, demographic and environmental crises. It is a source of problems: disease, migrants, difficult foreign policy choices. Robert Kaplan's imagery of future 'anarchy' and William Zartman's notion of 'collapsed states' mistakenly frame external perceptions of the subregion.[42] But few expect a future where West Africa will alter more than marginally its position as one of the poorest, weakest, most dependent, and politically unstable subregions of the global political economy, despite recent improvements in economic growth rates to an average of 4 per cent per annum.[43]

The purpose of this chapter has been to chart the efforts of ECOWAS to reduce this economic and political dependence and increase subregional cooperation. ECOWAS was created in 1975 to implement an ambitious subregional and regional economic integration agenda: as one of the elements of an African Economic Community, it sought to implement economic integration, initially by 2000. In practice it has been a means for Nigeria to assert itself politically and strategically within the subregion. Very little official economic integration has taken place although the unofficial, parallel economies of the states have inevitably become more integrated.[44] The timescale for economic integration, which has increasingly become the new open market regionalism, has slipped: to 2025. Over time, ECOWAS has changed in character, from being an agent of economic integration, to an institution devoted to more modest attempts at subregional diplomacy, conflict resolution and political cooperation, probably as an inevitable response to the subregion's growing political crises. ECOWAS is an example of subregionalism which is primarily state-promoted regional cooperation, with Nigeria's ambitions at its centre. The major problems in ECOWAS have thus been due to the diverse and frequently changing character of the political leaderships involved, the differing objectives of the states, the influence of external powers, such as France, and the collapse of over-ambitious, earlier state and subregional integration efforts. West Africa's growing economic, political and social crises, seen in the economic decline of the subregion and the effective implosion of states such as Liberia and Sierra Leone in the 1990s, have not been countered by a more assertive and effective subregionalism. West Africa's general lack of subregional cohesion has been accentuated by these growing crises.

In these circumstances, it is surprising that ECOWAS has managed to survive as an institution, given the markedly large gap between initial aspirations and achievements. It has survived despite the little progress it has made in its over-ambitious economic integration agenda. ECOWAS has not reduced, effectively managed, or coordinated the response to the growing marginalization of the subregion. Since 1975 there has only been a slow growth in intra-regional trade. There has been a great reluctance to reduce tariffs, because of the rentier state's need for income, and a massive growth in unrecorded or unofficial trade. ECOWAS has had very little economic impact, due to political controversies over 'illegal' immigrants, the lack of commitment by states (most are behind in paying their dues), and the politicization of the institution.

ECOWAS has also had virtually no impact representing the subregion, even in the 1990s when there has been a revival of interest in, and a recognition of the need for, a new regionalism in the increasingly globalized world order. ECOWAS in the late 1990s presents a remarkable paradox: it is a subregional organization founded for the purpose of economic integration, with few results. Yet it can be justified as a useful if at times problematic political-cum-security organization. Despite its exceptionally slow movement on official economic integration, ECOWAS has increasing relevance as a subregional security system in the context of a retreat from Africa by the major external powers. The now-common refrain, from many western and African diplomats and politicians, of 'an African solution to African problems' – in the political and security fields – provides a new rationale for a revitalized ECOWAS.

Such a subregional security system as ECOWAS has a number of uses in the late 1990s and into the next century. Despite its problems, the ECOMOG intervention eventually ended the Liberian civil war, and the subregional organization has had its uses in mediating border disputes and in diplomatic negotiations across old language and colonial lines. ECOWAS thus has had a relatively useful if narrow security function. But it now needs to concentrate upon the greater, more damaging new security issues as its already weak Westphalian pillars – the subregional state system – are increasingly undermined.

If we interpret ECOWAS as a state subregionalist project on the part of Nigeria, then Nigeria's gains have been relatively few compared to costs. Nigerian diplomatic successes are of some use internationally and play reasonably well to a Nigerian domestic audience. What is needed before ECOWAS can succeed as a subregional organization is for Nigeria to be transformed into a more democratic, less rent-seeking polity. State-led subregionalism in the hands of a less assertive, and more internationally respected state would help the subregion face the new challenges of hyper-globalization and could lead to greater cooperation on some pressing subregional issues such as debt, terms of trade, mass poverty, and basic human development.

NOTES

1. H.M.A. Onitiri, 'Changing Political and Economic Conditions for Regional Integration in SubSaharan Africa', and J. Fine and S. Yeo, 'Regional Integration in SubSaharan Africa: Dead End or Fresh Start?', both in A. Oyejide *et al.* (eds), *Regional Integration and Trade Liberalization in SubSaharan Africa* (Basingstoke: Macmillan, 1997). Unless stated otherwise, the statistics in this chapter are taken from the following: African Development Bank, *African Development Report 1997* (Oxford: Oxford University Press, 1997); UNDP, *Human Development Report 1997* (Oxford: Oxford University Press, 1997); and World Bank, *World Development Report: The State in a Changing World* (Oxford: Oxford University Press, 1997).
2. A. Hurrell, 'Explaining the Resurgence of Regionalism in World Politics', *Review of International Studies*, 21, 4, 1995, pp. 336–7.
3. Article 2(1) of the revised 1993 Treaty says that ECOWAS 'shall ultimately be the sole economic community in the region'.
4. J.E. Okolo, 'The Development and Structure of ECOWAS', in J.E. Okolo and S. Wright (eds), *West African Regional Cooperation and Development* (Boulder: Westview, 1990), pp. 20–21.
5. T. Abdul-Raheem (ed.), *PanAfricanism: Politics, Economy and Social Change in the Twenty-First Century* (London: Pluto, 1996), provides a reflection on these issues.
6. C. Lancaster, 'The Lagos Three: Economic Regionalism in Sub-Saharan Africa', in J.W. Harbeson and D. Rothchild (eds), *Africa in World Politics* (Boulder: Westview, 1991), p. 262; and C. Daddieh, 'Structural Adjustment Programmes and Regional Integration: Compatible or Mutually Exclusive?', in K. Mengisteab and I. Logan (eds), *Beyond Economic Liberalization in Africa* (London: Zed Books, 1995), p. 256.
7. S.P. Riley, 'The 1996 Presidential and Parliamentary Elections in Sierra Leone', *Electoral Studies*, 15, 4, 1996, pp. 537–44; and S.P. Riley, 'Sierra Leone: the Militariat Strikes Again', *Review of African Political Economy*, 24, 72, 1997, pp. 287–92.
8. D.C. Bach, 'Institutional Crisis and the Search for New Models', in R. Laverne (ed.) *Regional Integration and Cooperation in West Africa* (Trenton: Africa World Press, 1997), pp. 82–4, and Reuters, 30 August 1997.

9. F. Soderbaum, *Handbook of Regional Organizations in Africa* (Uppsala: Nordiska Afrikainstitutet, 1996), pp. 32–3; *Africa Confidential*, 12 September 1997; and PanAfrican News Agency (PANA), 2 September 1997.
10. Pompidou is quoted from J. Chipman, *French Power in Africa* (Oxford: Basil Blackwell, 1989), p. 127.
11. J. Fine and S. Yeo, 'Regional Integration in SubSaharan Africa: Dead End or Fresh Start?', in A. Oyejide *et al.* (eds), *Regional Integration and Trade Liberalization in SubSaharan Africa* (Basingstoke: Macmillan, 1997), pp. 428–76, and D. Cobham and P. Robson, 'Monetary Integration in Africa: a Deliberately European Perspective', *World Development*, 22, 3, 1994, pp. 285–99.
12. A. Bundu, 'ECOWAS and the Future of Regional Integration in West Africa', in R. Laverne (ed.), *Regional Integration and Cooperation in West Africa* (Trenton: Africa World Press, 1997), pp. 44–5, and ECOWAS, *Regional Peace and Stability: A Pre-Requisite for Integration*, Annual Report, 1993.
13. J.-F. Bayart, *The State in Africa* (Harlow: Longman, 1993), p. 253, and A. El-Kenz, 'Youth and Violence' in S. Ellis (ed.), *Africa Now: People, Policies and Institutions* (London: James Currey, 1996), pp. 42–57.
14. C. Leys, 'Confronting the African Tragedy', *New Left Review*, 204, 1994, p. 46.
15. J. De Mello and A. Panagariya, *The New Regionalism in Trade Policy* (Washington, DC: World Bank, 1992).
16. D.C. Bach, 'Institutional Crisis and the Search for New Models', p. 84, and A. Bundu, 'ECOWAS and the Future of Regional Integration in West Africa', p. 35, both in R. Laverne (ed.), *Regional Integration and Cooperation in West Africa* (Trenton: Africa World Press, 1997).
17. J.A. Wiseman (ed.), *Democracy and Political Change in Sub-Saharan Africa* (London: Routledge, 1995).
18. C. Daddieh, 'Structural Adjustment Programmes and Regional Integration: Compatible or Mutually Exclusive?', in K Mengisteab and I. Logan (eds), *Beyond Economic Liberalization in Africa* (London: Zed Books, 1995), p. 264.
19. The academic literature on this is enormous. A study which looks at the relationships between adjustment and the recent democratization of African societies, and contains a literature survey, is S.P. Riley and T.W. Parfitt, 'Economic Adjustment and Democratisation in Africa', in J. Walton and D. Seddon (eds), *Free Markets and Food Riots: the Politics of Global Adjustment* (Oxford: Blackwell, 1994), pp. 135–70.
20. K. Meagher, 'Informal Integration or Economic Subversion? Parallel Trade in West Africa', in R. Laverne (ed.), *Regional Integration and Cooperation in West Africa* (Trenton: Africa World Press, 1997), pp. 180–81, and H. Schissel, 'Africa's Underground Economy', *Africa Report*, 34, 1, 1989, pp. 43–6. This informal trade is not well-hidden or particularly secretive.
21. E. May, 'Exchange Controls and Parallel Market Economies in Sub-Saharan Africa: Focus on Ghana', World Bank *Working Paper* 711 (Washington, DC: World Bank, 1985), and R. Jeffries, 'Rawlings and the Political Economy of Underdevelopment in Ghana', *African Affairs*, 81, 384,1982, pp. 307–17, and P. Nugent, 'Educating Rawlings: the Evolution of Government Strategy toward Smuggling', in D. Rothchild (ed.), *Ghana: the Political Economy of Recovery* (Boulder: Lynne Reinner, 1991), pp. 69–84.

22. K. Meagher, 'Informal Integration or Economic Subversion? Parallel Trade in West Africa', in R. Laverne (ed.), *Regional Integration and Cooperation in West Africa* (Trenton: Africa World Press, 1997), pp. 177–8.
23. D.F. Luke and S.P. Riley, 'The Politics of Economic Decline in Sierra Leone', *Journal of Modern African Studies*, 27, 1, 1989, pp. 133–42, and D.F. Luke and S.P. Riley, 'Economic Decline and the New Reform Agenda in Africa', *IDPM Discussion Papers*, 28, University of Manchester, 1991. W. Reno, *Corruption and State Politics in Sierra Leone* (Cambridge: Cambridge University Press, 1995), is an analytically sophisticated and detailed account. S.P. Riley, 'The Land of Waving Palms': Corruption Inquiries, Political Economy and Politics in Sierra Leone', in M. Clarke (ed.), *Corruption: Causes, Consequences and Controls* (London: Frances Pinter, 1983), pp. 190–206, gives an indication of the early origins of the problems.
24. Some discussion of these issues is found in R.T. Naylor, *Hot Money and the Politics of Debt* (London: Unwin Hyman, 1987), R.T. Naylor, 'The Underworld of Gold', *Crime, Law and Social Change*, 25, 3, 1996, pp. 191–241, M. Hampton, *The Offshore Interface: Tax Havens in the Global Economy* (Basingstoke: Macmillan, 1996), and J. Robinson, *The Laundrymen* (London: Simon and Schuster, 1994).
25. B. Titley, *Dark Age: the Political Odyssey of Emperor Bokassa* (Liverpool: Liverpool University Press, 1997), p. 19.
26. J. Chipman, *French Power in Africa* (Oxford: Basil Blackwell, 1989), details the background. J.-F. Medard, 'The Underdeveloped State in Tropical Africa: Political Clientelism or Neo-patrimonialism?', in C. Clapham (ed.), *Private Patronage and Public Power* (London: Frances Pinter, 1982), p. 183, calls the Francophone elites a 'state bourgeoisie'. Other assessments of the close personal ties include G. Martin, 'The Historical, Economic and Political Bases of France's African Policy', *Journal of Modern African Studies*, 23, 2, 1985, pp. 189–208, and M. Staniland, 'Francophone Africa: the Enduring French Connection', *The Annals*, 489, 1987, especially pp. 53–5.
27. I.W. Zartman, 'Africa and the West: the French Connection', in B.E. Arlinghaus (ed.), *African Security Issues: Sovereignty, Stability and Solidarity* (Boulder: Westview, 1985), p. 67.
28. *Africa Confidential*, 18 July 1997. There are clear links between this 'rethink' and the preliminary skirmishes for the renegotiation of the Lome Convention, due to start in September 1998.
29. T. Forrest, *Politics and Economic Development in Nigeria* (Boulder: Westview, 1995), and S.O. Osoba, 'Corruption in Nigeria: Historical Perspectives', *Review of African Political Economy*, 23 (69), 1996, pp. 371–86. A recent account of the human rights record of the Abacha regime is: Amnesty International, 'Nigeria: No significant change – human rights violations continue', *AFR* 44/20/97, 22 September 1997.
30. A. Onadipe, 'Behind the Dark Glasses: a Portrait of General Sani Abacha', *International Relations*, 13, 4, 1997, pp. 69–78.
31. S.A. Olanrewaju and T. Falola, 'Development through Integration: the Politics and Problems of ECOWAS', in O. Akinrade and J.K. Barling (eds), *Economic Development in Africa* (London: Frances Pinter, 1987), pp. 69–70.
32. R.D. Kaplan, 'The Coming Anarchy?', *Atlantic Monthly*, February 1994, p. 60.

33. W. Reno, 'Reinvention of an African Patrimonial State: Charles Taylor's Liberia', *Third World Quarterly*, 16, 1, 1995, pp. 109–20, and W. Reno, 'Privatising War in Sierra Leone', *Current History*, May 1997, pp. 227–30.
34. S.P. Riley and M.A. Sesay, 'Sierra Leone: The Coming Anarchy?', *Review of African Political Economy*, 22, 63, 1995, pp. 121–6, and S.P. Riley, 'State Collapse and Social Reconstruction in Africa', *India Quarterly*, 24, 4, 1997, pp. 145–55. P. Richards, *Fighting for the Rain Forest: War, Youth and Resources in Sierra Leone* (London: James Currey, 1996), is a detailed examination of the shaky West African foundations of Kaplan's arguments.
35. J.A. Widner, 'States and Statelessness in Late 20th Century Africa', *Daedalus*, 124, 3, 1995, pp. 129–53, and J. Harding, 'The Mercenary Business', *Review of African Political Economy*, 24, 71, 1997, pp. 87–97.
36. J.E. Okolo, 'The Development and Structure of ECOWAS', in J.E. Okolo and S. Wright (eds), *West African Regional Cooperation and Development* (Boulder: Westview, 1990), p. 39, and M.A. Sesay, 'Civil War and Collective Intervention in Liberia', *Review of African Political Economy*, 23, 67, 1996, p. 43.
37. J.E. Okolo, 'Securing West Africa: the ECOWAS Defence Pact', *The World Today*, 39, 5, 1983, pp. 177–84.
38. V. Tanner, 'Liberia: railroading peace?', unpublished paper, December 1997, B. Ankomah, 'Taylor's Triumph', *New African*, December 1997, pp. 1–24, and *Africa Confidential*, 21 November 1997. Tom Ikimi, Nigeria's Foreign Minister, also claimed in November 1997 that the cost was US$4.0bn, but other sources suggest that the bill was nearer US$1.0bn.
39. The conflict and ECOMOG's role in it is discussed in S. Ellis, 'Liberia 1989–1994: a Study of Ethnic and Spiritual Violence', *African Affairs*, 94, 375, 1995, pp. 165–97; S.P. Riley, 'Liberia and Sierra Leone: Anarchy or Peace in West Africa?', *Conflict Studies*, 287, 1996, pp. 1–28; S.P. Riley and M.A. Sesay, 'Liberia: after Abuja', *Review of African Political Economy*, 23, 69, 1996, pp. 429–37, and Q. Outram, 'Cruel Wars and Safe Havens: Humanitarian Aid in Liberia 1989–1996', *Disasters*, 21, 3, 1997, pp. 189–205.
40. R.A. Mortimer, 'Senegal's Role in ECOMOG: the Francophone Dimension in the Liberian Crisis', *Journal of Modern African Studies*, 34, 2, 1996, pp. 293–306.
41. S.P. Riley and M.A. Sesay, 'Liberia: after Abuja', *Review of African Political Economy*, 23, 69, 1996, pp. 429–37.
42. R.D. Kaplan, 'The Coming Anarchy?', *Atlantic Monthly*, February 1994, pp. 44–76, and I.W. Zartman (ed.), *Collapsed States: The Disintegration and Restoration of Legitimate Authority* (Boulder: Lynne Reinner, 1995). Frank Furedi has recently identified a demonization of the South and a rethinking of the colonial experience. The post-colonial South becomes a danger to itself and the West. In addition, Mark Duffield describes how recent aid and development thinking has become focused around the issue of the containment or exclusion of threats to the West. The ideas of Kaplan and Zartman would seem to fit within this framework. F. Furedi, 'Moral Condemnation of the South', in C. Thomas and P. Wilkin (eds), *Globalization and the South* (Basingstoke: Macmillan, 1997), pp. 76–90, and M. Duffield, 'Complex Emergencies and the Crisis of Developmentalism', *IDS Bulletin*, 25, 4, 1994, pp. 37–45.
43. *West Africa*, 15 September 1997.
44. These aspects of contemporary West Africa feature largely in the expatriate fiction devoted to the subregion, such as Robert Wilson's two novels: *Instruments of Darkness* (London: HarperCollins, 1995) and *The Big Killing*

Part 2
Subregionalism in the Americas

Introduction

by Jean Grugel

'New' regionalism, meaning the emergence of state strategies for region-building based on market-led, investment-friendly development projects, with post-Cold War security issues also at the core, was identifiable most clearly and earliest in the Americas. The US made a quick response to the global transformation of 1989–90, re-engaging within its own hemisphere in a particularly energetic and imaginative fashion. The then President George Bush pushed region-wide open integration onto the agenda of inter-American relations in 1990 with the Enterprise for the Americas Initiative which proposed a free-trade community stretching from Anchorage to Tierra del Fuego. The most concrete achievement of this vision was the North American Free Trade Agreement (NAFTA) between the US, Canada and Mexico, taken at the time as the first stage of a hub and spoke pattern of integration to cross the Americas. The fundamental aim of NAFTA was to guarantee the US economy privileged access to the economies of the Western hemisphere and provide US capitalism with some security in the face of rising international competition. NAFTA, then, symbolized the return of the US to the Americas and a recognition on the part of the US state of its interdependent relationship with the rest of the Americas.

NAFTA ushered in a qualitatively new relationship between the US state and Mexico in a number of ways. However, in almost all aspects, the deal was profoundly asymmetric. What was perhaps surprising, therefore, was the eagerness which was displayed by the rest of the Latin American and the Caribbean (LAC) states, and even much of LAC capitalism, over the new regionalism. If the 'new' regionalism was essentially a way of defending and reorganizing US capitalism, why should LAC, traditionally the backyard of the US, greet the plan as a development option in which they were potential beneficiaries? The answer lies in the fact that the adjustments that all LAC states had either chosen or been forced to make in response to debt and recession in the 1980s and early 1990s had reshaped the political and economic choices of state elites. Throughout the LAC, nationalist development through import-substituting industrialization was abandoned and in its place a new economic orthodoxy of export promotion was gradually put in place, even in the small states of the Caribbean where the economies are intensely vulnerable to even relatively moderate fluctuations in external demand. The result was a LAC-wide consensus around the importance of foreign investment for growth and the centrality of the market for development. In this context, therefore, Americas-wide integration appeared at least to offer the prospect of reduced protectionism on the part of the US and the option of competing for investment from US transnational companies. For governments committed to open development and to preventing a return to protectionism, it also provided the opportunity to lock in the economic reforms, even when the costs to powerful domestic groups (for example, public employees, state companies and certain private business groups) were high. Consequently, the passing of NAFTA led to a flood of integration initiatives between the LAC states and the US and within the LAC itself, all couched in the language of open regionalism, aimed at securing tariff-free access to external markets.

Despite this enthusiasm, the project of regional wide integration centred on the US economy began to run into difficulties after 1992. On the one hand, domestic problems, especially opposition from Congress, and from labour and environmental activists, hindered the expansion of NAFTA southwards. On the other it rapidly became clear that Washington's chief concerns were actually centred on a small number of LAC states. From the US perspective, there might be no need to draw in all the LAC after all. In particular, the US state is concerned with managing the dense network of its political, security and economic interests in Mexico, with which it shares a long and highly permeable border. It also has an agenda of security concerns (essentially revolving around narcotics and immigration issues) and arguably a historical engagement with Central America and the Caribbean which it is difficult now to evade. But the extent of US commitment to states south of Venezuela and Colombia is domestically contested and therefore much more fluctuating in depth and intensity.

Yet, as the project of region-wide integration under the auspices of the US state faltered, subregionalism within the LAC gathered steam. The LAC states took up the challenge of designing new patterns of subregional integration which conformed to the patterns of open development and kept the door open to US engagement. Two of the chapters in this section examine the most significant and developed subregional integration projects: the Common Market of the South (MERCOSUR), linking Brazil, Argentina, Uruguay and Paraguay, and the Association of Caribbean States (ACS), a cross-cultural region-building project that tries to bring together the small states of the Caribbean and Central America. Other initiatives for subregional integration include the Group of Three (Venezuela, Colombia and Mexico) and the revival of the Central American Common Market. The third chapter in this section examines in

Introduction: Subregionalism in the Americas 93

depth the emergence of a somewhat different approach to subregional integration, that of Chile, which has sought to deepen its ties with both NATFA and the MERCOSUR bloc of states, without accepting either the political, security or environmental clauses which come with full membership.

The resurgence of subregionalism throughout the LAC, as the participation of the US became problematic, demands an explanation. In part, this must necessarily be economic. Subregional integration remains a defensive response on the part of the LAC to its economic marginalization in the 1980s. It acts as a vehicle for increasing the connections between the LAC and the global economy. Moreover, even without the enticement of US markets, the LAC states are now constrained to pursue open development by the very weight of the liberal economic model which is now in place. New export markets are central to any project of liberal economic growth. And for Brazil, Argentina, Chile, the LAC markets are important destinations for a range of nationally produced goods. But it would be wrong to suggest that subregional integration in the LAC is merely an economic project. Part of its resilience, in fact, lies in its political, cultural and historic foundations with the area. The language of political elites of almost all the LAC states, for example, still echoes the grandiose dreams set out with the Liberation from Spain in the nineteenth century, when Independence was debated in terms of presenting a Latin bloc in opposition to, not in collaboration with, the US. More prosaically, contemporary identity politics of the region is shaped by a common commitment to at least formal democratization and political cooperation. Also, the 'new' regionalism arises out of the ashes of old regionalist integration schemes promoted, amongst others, by the Economic Commission for Latin America and the Caribbean (ECLAC), a highly influential UN-funded think tank throughout the region. Although the 1960s integration projects, which ranged from the Latin American Free Trade Association and the Central American Common Market, both established in 1961, to the Andean Pact and the Caribbean Community created in 1969 and 1973 respectively, were all condemned as failures, they nonetheless contributed to building a basis for political cooperation to a common sense of belonging. None of this means, of course, that subregionalist schemes are without tensions, rivalries and internal problems. Many of these are explored in the chapters that follow. But it does signify that subregionalist integration schemes are culturally and politically embedded and that there is a real commitment on the part of governing elites to making at least aspects of them work.

5 MERCOSUR: From Domestic Concerns to Regional Influence

Paul Cammack

The origins of MERCOSUR (*Mercosul* in Portuguese, the language of Brazil) lie in the bilateral discussions between Argentina and Brazil which led to the signing of a cooperation agreement between Presidents Raúl Alfonsín and José Sarney respectively in 1986. The incorporation of neighbouring Paraguay and Uruguay into continuing negotiations led to the signing of the Treaty of Asunción (this being the capital of Paraguay) between the four South American states in March 1991. The chief protagonists on this occasion were presidents Carlos Menem of Argentina, Fernando Collor de Melo of Brazil, General Andres Rodriguez of Paraguay, and Luis Alberto Lacalle of Uruguay. This was followed by agreement in 1992 on a customs union which came into force on target in January 1995, in conjunction with a common external tariff.[1] By this point, the return of the states of the subregion to democratic rule was complete, following the inauguration of President Carlos Wasmosy in Paraguay in August 1993. Despite the entry of Paraguay and Uruguay, the subregional grouping is dominated by a single state, Brazil, and virtually constituted, in terms of size, by Brazil and Argentina alone. As of 1990, Brazil alone accounted for 79 per cent of the population to be incorporated into MERCOSUR (149 of 189 million), and 71 per cent of the total GDP ($333bn of $470bn). Argentina and Brazil together accounted for an overwhelming 97 per cent of the total population, and 96 per cent of the total GDP.[2] The manner of its emergence and the balance between its members suggest that MERCOSUR should be seen as a bilateral venture between Brazil and Argentina, which Paraguay and Uruguay have felt constrained to join, rather than as a common union between four equal partners.

Although the completion of the common market (with the phased removal of exceptions for particular products and exemptions for specially protected free-trade zones) will not be achieved fully for two further decades, the phase of subregional institution-building which took place in a short space of time in the early 1990s was significant and in some ways surprising: significant because of the past history of rivalry and hostility between the states concerned (particularly between Argentina and Brazil), and surprising because of the unequivocal commitment of the association to 'open regionalism' and to the promotion of free trade on broadly neo-liberal principles. Previous regional associations had sought to promote strategic rather than free trade, and to reinforce and stimulate protectionist domestic regimes; and Brazil in particular had always been committed to projects of state-led economic development which relied upon protection and import-substitution rather than upon the promotion of exports through the liberalization of trade. This shift of orientation can be understood in terms of the changed global and regional (Latin American) context. But it is a shift that is entirely in harmony with larger processes taking place in the Americas, and in particular with the growing regional commitment to neo-liberal economic principles and liberal democracy. The forces driving the creation of MERCOSUR have not reflected a subregional effort to establish a new and separate identity in opposition to US hegemony in the region as a whole, as is the case with the EAEC (see Chapter 10).

On the contrary, commitment to the association has derived from the logic of projects of domestic restructuring intended to bring its members into line with the emerging regional pattern: the subregional project has reflected and supported the efforts of Argentina and Brazil (and to a lesser extent Paraguay and Uruguay) to catch up with their neighbours by implementing and consolidating domestic projects which aim to establish a neo-liberal economy capable of being sustained through liberal democratic institutions. Because this is so, the project should be understood as much in terms of domestic political economy (in other words, class interests and state projects for capitalist accumulation and legitimation at the national level) as in terms of a 'new international political economy' focused on the subregional, regional or global level. MERCOSUR, as a subregional project, is as much about creating and securing a particular kind of domestic order as about establishing state positions in an emerging world order, and above all, it has sought to reinforce the reorganization of national space along particular political and economic lines. Engagement in the subregional project, in short, has been intended to advance and reinforce emergent national political projects whose principal goal is to achieve bourgeois hegemony and relative state autonomy at the domestic level, and specifically to transform social relations in order to embed a neo-liberal domestic order which it is possible to reproduce through liberal democratic political institutions: in this sense, the creation of the subregional association can be seen as an extension of domestic political economy. Once in being, however, it is beginning to exert an influence at regional level: NAFTA and MERCOSUR are now seen as the building blocks of a larger regional free-trade area of the future, and there are signs that the consolidation of MERCOSUR has produced a counterweight in the region to the hegemony of the United States, and facilitated the emergence of a new agenda. It may be, therefore, that it will eventually bring about a shift of power within the region.

In both Brazil and Argentina, it has proven extremely difficult, under military and civilian rule alike, for successive governments to impose on either capital or labour the disciplines which successful capitalist reproduction requires. Equally, it has proven difficult to find a basis on which to legitimize successive political regime forms. This became critical in the 1990s, as a regional consensus formed among governing elites on the need for a decisive break with the past and on the superiority of neo-liberal economic regimes sustained by liberal democratic politics. The creation of MERCOSUR, and beyond it the construction of a free-trade area across the whole of the Americas, is primarily intended to lock each country into the requisite forms of social and economic discipline, and thereby make it possible both to impose the desired neo-liberal economic regime, and to maintain it through liberal democratic means. Within this common framework, of course, states compete with each other, and the members of MERCOSUR naturally seek to use it to assert their own priorities over those of a large economic rival, the United States.

The strengthening of bilateral ties between Argentina and Brazil in the mid-1980s took place in the immediate context of a recent return to democracy in each case, and at its inception the process could be considered as a limited venture primarily intended to provide support for democratization.[3] At one point, too, progress towards a larger union might have reflected a defensive move in reaction to the creation of NAFTA, aimed at creating an enlarged space for trade and investment in South America. But as more comprehensive national projects of liberalization and democratization took shape in the two countries, the subregional initiative broadened in scope, and became increasingly shaped by considerations of domestic political economy in a new global context.

The key aspect of the regional context here was the final disappearance of the conditions for statist developmental projects (based on import-substitution) as a result of the debt crisis of the 1980s. This was made critical by the relative backwardness of both domestic adjustment and political institutionalization in Argentina and Brazil. These two countries (accustomed to seeing themselves as the leading economies of South America) had made little progress in restructuring their economies by 1990, and their processes of transition to democracy (following military withdrawal in 1983 and 1985 respectively) remained precarious and only weakly institutionalized.

While the initiation of bilateral agreements coincided with the return to democracy, the genesis and consolidation of MERCOSUR coincided in time with the adoption of programmes of neo-liberal restructuring in Brazil in 1990 (under President Fernando Collor de Melo) and in Argentina in 1991 (under President Carlos Saúl Menem). In each case, they came against the background of repeated failures in the recent past. And in the case of Brazil, the failure of Collor's project and the delay in successful stabilization until Cardoso's currency reform (known as the *real* plan, the *real* being the new currency) took hold in 1994 only exacerbated the sense of urgency, and of the need for external support. Hence the utility of seeing the building of MERCOSUR as an auxiliary means of promoting and guaranteeing neo-liberal restructuring and liberal democratization at national level in each case. The focus here, therefore, will be on the emergence of broadly parallel national projects in Argentina and Brazil. After a discussion of the regional context, the economic, political, ideological and security dimensions of MERCOSUR are explored for each. In view of the intimate association between political and ideological concerns, these themes are treated together. Thereafter, the creation of the four-state association is reviewed in the context provided, with subsidiary attention to the cases of Paraguay and Uruguay. In conclusion, an assessment will be made of the mutually reinforcing dynamics of the domestic, sub-regional and regional dimensions of the association.

THE REGIONAL CONTEXT

The immediate regional context behind MERCOSUR is set out by Payne and Grugel in the previously published companion collection to the present one.[4] In US–Latin American relations the key moves were the establishment of a US–Canada free-trade area in 1989, the first approaches from Mexico to the US in March 1990, the launching by President Bush of the Enterprise for the Americas Initiative in June of the same year, and the signing of the NAFTA agreement in August 1992, to come into force after ratification in the signatory countries in January 1994. Framework agreements with other Latin American states (including the MERCOSUR Four) were designed 'to set out and enforce new economic and political "rules of the game" in the hemisphere,' reflecting 'the triumph of economic liberalism, of faith in export-led growth and of belief in the centrality of the private sector to the development process'.[5] As Payne notes, the proposal for a series of free-trade agreements which might eventually create a hemisphere-wide free-trade association stretching from Alaska to Antarctica was received enthusiastically precisely because of 'the broad shift from import-substitution to export-oriented strategies of development which had taken place across [Latin America] since the beginning of the 1980s'.[6] Grugel traces this process back to the Mexican default of 1982 and the consolidation of export promotion and deregulation across the region thereafter, and relates it to the simultaneous process of democratization throughout the decade. Overall, she suggests, development models had to be rethought, but could no longer be generated and sustained from within. As a result, 'external prescriptions were sought, principally in the shape of assistance from multilateral donors and the US, the effect of which was to impose orthodox solutions on the region'.[7] Bilateral agreements proliferated in the early 1990s, and in December 1994, at the First Summit of the Americas in Miami, the states of the region (with the exception of Cuba) committed themselves to negotiations towards a Free Trade Area for the Americas, and proposed to launch the negotiating process at a second summit, to be held in Chile in 1998. By the early 1990s, a coherent project of open regionalism had taken shape in the region. As presented by the influential UN Economic Commission for Latin America and the Caribbean (ECLAC) its overall goal was to promote international competitiveness through the attraction of foreign investment and technology. It would build upon trade liberalization and deregulation processes at national level; it would reinforce the efficiency and coherence of policy-making at that level; it would ideally promote a regional or hemispheric free-trade area in an open global economy; and if the global economy turned away from open trading, it would at least preserve an expanded regional market which would offer compensation for protectionism elsewhere.[8]

Two cases were influential in the region – those of Mexico and Chile. Domestic economic liberalization was pursued with vigour following the assumption of the Mexican presidency by Carlos Salinas in 1988, and was well underway prior to the opening of the negotiations which eventually led to the NAFTA agreement, while Chile's liberal economic model had been in place for 15 years and had become acceptable to the leading civilian political forces by 1990, the year of the transition from Pinochet's rule. In discursive terms, Chile's transition to democracy was the most significant development in the region, as it paved the way for the celebration of the Chilean model, and even the retrospective endorsement of the manner in which it was put into place. The smooth restoration of pro-market centre-party politics in 1990 created precisely the commitment to the new rules of the economic game in a climate of democratic accountability that the ideologues in the National Endowment for Democracy were busily promoting in the region.⁹ As a result, the organic intellectuals of the US regime and its new hemispheric policy were quick to seize on the Chilean case. The climate of the time is reflected in a volume published by the Inter-American Dialogue and supported by both the National Endowment for Democracy and the A.W. Mellon Foundation, in which Scully argued that 'In many ways, Chile has been extraordinarily fortunate ... Strong political institutions arising from the democratic past, made in part more conducive to consensus policy-making by authoritarian holdovers built into the 1980 Constitution, have endowed Chile's democratic government with a remarkable capacity to implement and sustain coherent economic policy.'¹⁰ Brazil and Argentina suffer in comparison, despite their earlier transitions to democracy – Argentina is described as a case of 'democracy in turmoil', while Brazil is diagnosed as suffering from 'the hyperactive paralysis syndrome'. In Argentina, de Riz describes a situation in 1990, prior to the economic reform which eventually succeeded in stabilizing the currency, of hyperinflation and extreme political instability; in Brazil Lamounier identifies, from the mid-1980s to the early 1990s, 'a syndrome of declining governability rooted in a pervasive feeling, among the country's elites, of insecurity regarding their cohesion and legitimacy and aggravated by a mistaken attempt on their part to meet the problem by constantly overloading the formal political agenda'.¹¹

Overall, therefore, Argentina and Brazil were seen as losing ground in an emerging pattern of economic reorientation and political transition, in an ideological climate entirely dominated by the ideals of neo-liberalism and liberal democracy. In this context, security issues were redefined in two ways: with an emphasis on the negative consequences of any potential for renewed military intervention, and on the positive need for economic stability and progress as a means of promoting security in the future. Over the period from the mid-1980s to the early 1990s national projects took shape in Argentina and Brazil which were congruent with this emerging regional pattern, and commitment to MERCOSUR flowed from the shape and logic of these national projects.

NATIONAL PROJECTS AND SUBREGIONALISM IN ARGENTINA AND BRAZIL

The twentieth-century political economy of Argentina and Brazil has been dominated by three successive national projects. The first – export orientation – centred on the exchange of primary commodities for manufactured goods from abroad, and developed under elite civilian rule from the late nineteenth century onwards. The second – ISI or import-substituting industrialization – centred on inward-oriented national development aimed at developing the domestic market, and developed under populism from the 1940s onwards. The third – associated-dependent development – promoted industrial exports by attracting investment from multinational corporations, and developed under 'bureaucratic-authoritarian' military rule.[12] The first phase took place in Argentina under a centralized oligarchy established in 1880, and expanded between 1912 and 1916 to incorporate the (male) middle class into the electorate; in Brazil, it took place under a federal republic (established in 1889) in which power remained decentralized and highly oligarchic until 1930. The second phase took place in Argentina under the populist regime of Juan Domingo Perón (1946–55), within a 'developmentalist alliance' in which an interventionist state sought support from state-controlled labour unions and a state-backed national bourgeoisie. In Brazil Getulio Vargas put together in the early 1940s a complex developmentalist alliance which reached its peak under Juscelino Kubitschek (1955–60). The third phase was inaugurated first in Brazil, where the military intervened in 1964 and remained in power until 1985, seeking to exclude the organized working class and forge a new alliance between domestic and international capital. It was pursued in Argentina under successive periods of military rule between 1966 and 1973, and 1976 and 1983, interrupted by the return of a Peronist regime between 1973 and 1976. In each case, it was the failure of the military project which prompted a return to democracy, and the contemporary regional and global context which prompted the eventual adoption of a neo-liberal economic project.

The economic dimension of subregionalism

Despite their repressive character (particularly acute in the case of Argentina) neither military regime abandoned the populist legacy nor established a more liberal economy; neither did they forge ahead with reducing the role of the state. In the case of Argentina, intense social conflict combined with division among elites and among the military themselves meant that no clear new state project was consistently pursued: on two occasions, 1966–70 and 1976–80, a single military leader remained in office for four years, but in each case a nascent national project broke down. In the early 1970s the military recalled Perón in desperation, and with desperate consequences – Perón himself died within a year, and under his second wife, Isabel Perón, the country descended into conflict and chaos. In the early 1980s, rising social conflict led to military adventure, and the regime was brought down by the ill-advised invasion of the Malvinas/Falklands in 1982. Neo-liberal policies had been followed under Videla between 1976 and 1981, although at the same time the wide-ranging industrial activities connected with the military themselves had been protected, but the programme was abandoned once he left office, and its effects were obliterated by the effects of the Malvinas adventure. When the newly elected President Raúl Alfonsín took office in 1983 he was faced with a legacy of hyperinflation, economic stagnation and spiralling foreign debt. This he sought to address by means of two successive 'heterodox' economic stabilization programmes – the Austral Plan of June 1985 and the second Austral Plan of February 1987. Neither succeeded, and he left office in 1989 with a bleak economic record: GDP had fallen by over 3 per cent in 1988, inflation was peaking at just under 80 per cent per month (and would pass 3,000 per cent for the year), the balance of payments was $9bn in deficit, and the foreign debt had risen by 63 per cent to $65bn. As a result, Argentina's relations with the International Monetary Fund, the World Bank and creditor banks became extremely strained, and in mid-1989, just before Menem assumed power, World Bank finance was suspended. Menem began his period of office, therefore, in the midst of an acute economic crisis in which relations with the major global funding institutions were extremely poor.

In the case of Brazil, a rather different trajectory led to a similar position at the end of the 1980s. The Brazilian military, in power for 21 years, showed more continuity and achieved more success in economic policy than was the case in Argentina. However, its period of greatest success, featuring high rates of growth between the late 1960s and the 1970s, was based upon an intensification of the state-interventionist policies of the populist period, rather than upon a turn towards a more neo-liberal approach. A strategy based upon heavy state investment, particularly in such strategic areas as energy, heavy industry, and capital goods and arms production, was then intensified rather than abandoned after the first oil crisis: the Second Development Plan of 1974, implemented by General Ernesto Geisel, was a deliberate attempt to use a counter-cyclical policy of large-scale state investment to keep growth moving forward in adverse global conditions, and its failure left the military with an economic crisis which intensified in the period of drift after 1979.[13] A severe economic recession between 1980 and 1983 wiped out many previous gains, and led to a handover of power to civilian President José Sarney in 1985 in conditions of weak economic recovery, rising inflation, and foreign debt. Sarney's heterodox Cruzado Plan, introduced in February 1986, was only briefly successful in halting inflation, and a second attempt, through the 'Summer Plan' of 1987, also failed. Severe economic recession followed, and Sarney's term ended in 1990 with negative growth (with GDP falling by 4.1 per cent in 1990), hyperinflation (rising to over 3000 per cent in 1990), a large deficit on the balance of payments (over $12bn in 1989) and high foreign debt ($110bn in 1989). As in the Argentine case, relations with international financial institutions proved difficult during this period, with Brazil announcing a unilateral moratorium on $68bn of debt in 1987, and being denied a renegotiated stand-by IMF loan in 1989 after new terms were again not met.

In both Argentina and Brazil, then, the new regimes which took over at the end of the 1980s (Peronist Menem in Argentina, and the independent Collor de Melo in Brazil) faced conditions of severe economic instability. Each launched programmes of economic liberalization, which combined renewed currency reform with tariff reductions and privatization. In Argentina, a privatization programme was launched in 1990, a series of measures were taken to undermine job stability, and, after an unsuccessful initial experiment, the appointment of Domingo Cavallo as Minister of the Economy in January 1991 was followed by the linking of the *austral* to the US dollar, and then by the abandonment of the ill-fated *austral*, and a return to a new (and newly stable) *peso*.[14] In Brazil, a programme of tariff reduction (from 1990) and privatization (from 1991) began to construct Collor de Melo's proposed 'New Brazil', but inflation hovered at 450–550 per cent per annum through 1991 and 1992, and the prospects for stabilization were severely jolted by the impeachment and removal from office of President Collor on charges of corruption in 1992. Most significantly, a report published in 1991 by the Argentine Foreign Ministry identified the lack of 'labour flexibility' as the pressing problem in both Argentina and Brazil.[15] On all counts, the two were badly placed to compete with their neighbours.

In the context described above, the establishment of MERCOSUR in the early 1990s was clearly aligned with a long-delayed comprehensive programme of domestic economic reform which sought to promote and consolidate economic liberalization. In contrast, the bilateral accords of the mid-1980s had been accompanied only by half-hearted attempts at stabilization, without initiatives in the areas of tariff reduction and privatization. The value of MERCOSUR, from the point of view of Argentina and Brazil, was that it would advance and consolidate domestic reform. At the same time, the conversion of these two large South American economies to principles of neo-liberal economic management already entrenched in the region would remove a significant obstacle to a commitment to open regionalism 'from Alaska to Antarctica'. Without Brazil and Argentina, there could be no meaningful regional free-trade area, but once these two major South American economies were committed to a common programme of liberalization, any subregional association of which they were members would be influential. From this point on, therefore, domestic, subregional and regional dynamics might be mutually reinforcing.

The political and ideological dimensions of subregionalism

Priority was attached to the creation of MERCOSUR in Argentina and Brazil as a consequence of the launching of neo-liberal economic projects at the end of the 1980s. As seen in the previous section, the economic context was made critical by a combination of long-term and short-term factors: over the long term each country had pursued a pattern of predominantly state-led development since the Second World War, and an alternative liberal model of development had proved unsustainable when it had been attempted. In the short term, each government had found itself in the grip of a deepening economic crisis, and felt itself to be losing ground, in an increasingly competitive global environment, against rivals who had apparently adjusted to new circumstances with greater success. The depth of the political and institutional crisis in each country was if anything greater. In the long-term perspective, civilian and military liberals alike had proven politically weak throughout the period. In particular, no political party with an agenda of economic liberalism enjoyed success in either country after 1930. When liberal projects were devised in the late 1980s, therefore, there was not only no immediate party political support for them, but virtually no plausible political vehicle for them in living memory. Crucially, in each case, the hold of the regime of organized labour was precarious. In Argentina, the combative legacy of Peronism had not been extinguished, while in Brazil the emergence of the PT (Workers' Party) after 1979 had created the first independent mass party of labour.[16]

In these circumstances, each of the political regimes under which reform was advanced was fragile. In Argentina, Carlos Menem was elected by the Peronists – the political tradition defined by labour militancy and opposition to a liberal orientation in economic affairs. In Brazil, Fernando Collor de Melo was elected by a coalition united only by opposition to 'Lula' – Luis Inacio da Silva, the leader of the Workers' Party – and was backed initially only by a minuscule party of his own creation. In summary, a long-term crisis of political liberalism was resolved by carrying liberal projects forward on the basis of personal projects which lacked committed institutional or ideological support. This greatly reinforced commitment to MERCOSUR. Menem and Collor had strong motivation to offer each other mutual support, and to seek to strengthen commitment to the liberal project by furthering subregional cooperation. It is essential, therefore, that the depth of the historical crisis of political liberalism in each country is appreciated.

The periods of extended military intervention inaugurated in Argentina in 1966 and in Brazil in 1964 marked the culmination of a crisis of political institutionalization which dated back to 1930. In that year, the onset of global depression precipitated a collapse of political regimes which were already suffering intensifying internal strains. In Argentina, Radical Party rule inaugurated in 1916 and increasingly sustained by the political distribution of state resources through metropolitan and provincial clientelistic machines came to an end with the overthrow of the Irigoyen regime, and the installation of the dictatorship of General Uriburu.[17] In Brazil, the much more narrowly based oligarchic republic sustained by allied state elites broke down under the impact of dissent between the elites themselves, and a succession of challenges from below. Getulio Vargas, coming to power in October 1930 at the head of a provisional government after leading a rebellion which challenged his defeat in presidential elections, was to dominate politics through to the time of his suicide in 1954.[18]

In due course, the politics of populism came in both cases to offer a 'second-best' solution to the problems caused by a loss of economic direction and of the political hegemony of ruling elites. At the same time, populism was both anti-liberal and anti-democratic. It combined the promotion of state-led industrial development in the face of shrinking overseas markets and rampant protectionism in the international economy with the construction of political regimes built upon the corporatist inclusion of labour in the developmentalist alliance. In Argentina Perón acted as Vice-President and Secretary of Labour in the military government in power from 1943 to 1946, and from those positions built up a state-controlled labour movement which remained the bastion of Peronism through to the 1980s. He won election to the presidency successively in 1946 and 1951, supported by the Peronist (officially, the Justicialist) Party, and was then overthrown and exiled by military intervention. In Brazil, Vargas was the head successively of a provisional government (1930–34), a constitutional regime (1934–37) and a military-backed dictatorship (1937–45) before being forced from power at the end of the Second World War. By this time he too had set up a system of state-managed labour unions, and created not one but two political parties – the PSD (Social Democratic Party), based on state-level pro-regime elites, and the PTB (Brazilian Labour Party), based on the recently established system of labour unions. He returned to power in 1951 with their support, as did successive presidents (except for a brief interlude in 1960–61) until the military intervened in 1964.[19]

In both cases, populism was created from within what might be described as modernizing or developmental dictatorships, influenced by, though not directly comparable with, European fascism. It combined certain progressive themes – national developmentalism, the selective inclusion of hitherto marginalized sectors – with the systematic use of state resources to reward supporters and punish opponents. It thereby placed virtually insuperable obstacles in the way of the emergence of liberal democratic forms of politics. In both cases, the succeeding military regimes sought to construct a civilian alternative to the populist coalition, and in both cases they failed. The legacy of populism followed by extended military rule was an enduring crisis of political institutions, acute at the point of return to civilian rule (in 1983 in Argentina and in 1985 in Brazil), and continuing through the 1980s.

In Argentina, the military three times allowed the restoration of civilian rule after 1955, twice excluding Peronism and thereby permitting the election of leaders of different wings of the now divided Radical Party (Frondizi, 1958–62; Illia, 1963–66), and finally acquiescing in the inevitable arithmetic logic of Argentine politics and allowing the return of Perón (1973–76). On their departure in 1983 the electorate continued to be divided between the Radicals (only recently reunited) and the Peronists (badly divided after the latest disastrous experience of government). The failure of Alfonsín (Radical President, 1983–89) to restore economic order created an intense political crisis which was resolved in as surprising a manner as could be conceived – the election of a Peronist president who immediately launched a new political and economic project which repudiated virtually everything for which Peronism historically stood. President Menem, identified at the time of his emergence as a candidate with the opponents of modernization and democratization in the Peronist camp, immediately championed the shock programme of economic liberalization outlined above, in the face of opposition from much of the Peronist Party and the pro-Peronist unions.

In Brazil, relations between the military and civilian political elites took the form not of a pattern of alternation in power but of the creation of a hybrid regime: the military remained continuously in power from 1964 to 1985, but with brief exceptions they maintained a controlled regime of civilian parties and of national legislative elections.[20] Executive power remained firmly in military hands, giving the 'governing' party (set up under the name of ARENA (National Renovation Alliance) in 1965, and renamed the PDS (Social Democratic Party) in 1979, a purely electoral role. Strict rules of party membership and organization, reinforced by the systematic use of state resources to buy votes for the regime, preserved an increasingly precarious pro-government majority until the late 1970s. The steady electoral advance of the opposition MDB (Brazilian Democratic Movement) prompted the restoration of multi-party politics in 1979. Thereafter the military lost control successively over Congress itself and over the Electoral College, with the result that opposition candidate Tancredo Neves, was elected to the presidency in 1984.

In the meantime, however, the shift of substantial numbers of the most clientelistic politicians in the governing party over to the opposition made the renamed PMDB (Party of the Brazilian Democratic Movement) a far more heterogeneous and conservative party than it had been a decade before. The overall coherence of the party-political system was further damaged by the critical illness of Neves on the eve of his scheduled inauguration in March 1985. He was to die in April of the same year, and was replaced by the vice-presidential candidate, José Sarney, who until 1983 had been the head of the pro-military PDS, and had left it as a result of internal splits in return for the vice-presidential slot on the opposition electoral ticket. This combination of circumstances blocked the process of political renewal, destroyed both major parties emerging from the period of military rule, and gave an enlarged national role (particularly in two successive presidential elections) to the newly established PT (Workers' Party) led by independent trade unionist Luis Inacio 'Lula' da Silva. A protracted process produced a new constitution in 1988, but at the same time completed the destruction of the PMDB, prompting the radical intellectual and São Paulo senator Fernando Henrique Cardoso to lead a small group of committed Social Democrats in the party into the PSDB (Brazilian Social Democratic Party). In the meantime, Sarney's unpopularity, swelled by intensifying opposition arising after the failure of the 1986 stabilization programme, boosted support for the Workers' Party. Its leader, 'Lula,' along with the populist Leonel Brizola, became identified as by far the strongest presidential contenders as the first national elections since the 1950s approached. The parties of the centre and right were in total disarray, and there was no identifiable liberal candidate with good electoral prospects, or any party political vehicle which could sustain such a candidate.[21]

This was, as noted at the beginning of this section, the outcome of the ideological hegemony of developmentalist, corporatist and interventionist ideas, and the long-run weakness of liberal ideas in each country since 1930. It should be noted, in this respect, that in both countries liberalism had proved equally weak in civilian party politics and within the military.

In Argentina, liberal civilian traditions have been virtually excluded from power since the 1940s by the electoral weight of Peronism, surviving as a strand within or on the margins of politics. At the same time, the military has been split between liberal and corporatist factions, with the latter tending to predominate: an avowedly liberal project promoted by General Lonardi immediately following the exiling of Perón in 1955 was swiftly abandoned with his replacement by Aramburu in 1956, and only Videla thereafter sustained for long what was a broadly neo-liberal project. In Brazil, the uncompromisingly liberal UDN (National Democratic Union) played the role of a minority opposition from its founding in 1945 (confounding hopes that it would come to power in the wake of the ousting of Vargas) through to 1960, when it achieved a brief foothold on power by supporting the independent Janio Quadros (1960–61) in his successful bid for the presidency. His abrupt resignation after seven months in power, partly precipitated by his failure to develop a working relationship with the party, hastened its descent into incoherence. In the military, too, a recognizably liberal orientation was briefly espoused by the first post-coup president, General Humberto Castello Branco, but by the time he stepped down in 1967 the proponents of a corporatist and developmentalist approach had the upper hand, and the next military president to emerge from the *castelista* faction, General Ernesto Geisel, accentuated the statist and developmentalist aspect of the military project.

Liberalism as an ideological project, therefore, enjoyed little institutional support in either Argentina or Brazil at the end of the 1980s. Menem won power as a consequence of the failure of the Alfonsín government to address the economic crisis, but did so on the basis of the support of the historical opponents of liberalism. Collor campaigned on a programme of liberal modernization, threatening to purge corrupt political elites and cut back the bloated ranks of state employees maintained for clientelistic purposes, but he won because all factions of the elites rallied to him in the absence of a more credible candidate to oppose to Lula's socialism and Brizola's populism. Each came to power convinced that half a century of national history had to be thrown into reverse. It is hardly surprising that each sought what support they could not only from each other, but also from a political project aimed at building an enlarged common market in South America. With the fall of Collor, there was an even greater need for external support for democracy in Brazil.[22]

At this point involvement in MERCOSUR still reflected a domestic logic for both Argentina and Brazil, but it was a more comprehensive one than had operated when a bilateral alliance had been established in the mid-1980s. While the principal political concern then had been to provide support for a project of democratization, it was now to provide support for specific political projects committed to the introduction and consolidation of systematic programmes of neo-liberal economic reform. Even so, the project of democratization itself was not yet secure in either case, and concerns regarding military threats to democracy continued to play a prominent part in the subregional project.

The security dimension of subregionalism

The political and economic aspects of the creation of MERCOSUR, set in the broad ideological context of renewed regional and national commitment to neo-liberalism and liberal democracy, have been more significant than its security aspects. In addition, the significance of the security aspects, such as it was, have diminished over time. If they carried primary weight at any point, it was during the 1980s, when the military had only recently stepped down from power. At this time bilateral concerns were paramount, and the broader projects reviewed above had not yet taken final shape. In comparative terms, the key feature of security concerns in the region was the threat that the military might seek a return to power: there was and is no other significant regional or inter-regional security issue to parallel those which might be found in other regions of the world.[23] Nevertheless, the security dimension of MERCOSUR should not be neglected. In addition, there was a substantial history of economic and military rivalry between Brazil and Argentina which had to be overcome if the process of integration was to succeed.

The Brazilian military government espoused a regional geopolitical project focused on the idea of expansion into the interior, and the exertion of influence over its neighbours, and although this largely drew their attention to Amazonia it was erected upon a foundation of historical antagonism towards Argentina. At the same time, a protracted dispute with Chile over the Beagle Channel to the south which came to a head in the final period of military rule, in addition to the Malvinas adventure, suggested that the Argentine military had expansionist ideas of their own. More directly, each state under military rule had appeared to have rival ambitions in the sphere of nuclear power. Such aspects of bilateral negotiation as the cooperation between the two states on issues relating to nuclear technology should therefore not be discounted.

Nevertheless, the predominant security concern of civilian leaders in both countries (and in Paraguay at least at a later stage) was to prevent the military from returning to power. Given the identification of the military in each case with developmentalist and corporatist forms of politics, the adoption of liberal projects and the consolidation of those projects through subregional association could be regarded as having a security dimension, and being aimed both at a repudiation of past military intervention, and a shield against intervention in the future. For three reasons, the prospect of renewed military intervention was seen as a threat to future development in both Brazil and Argentina. First, military control over the process of economic development was seen as a threat to consistent and effective policy-making, especially in the light of the adventurism of the Argentine military, but also as a consequence of the misjudged expansionism and subsequent paralysis of economic management in Brazil since the late 1970s. Second, the presence of corporatist and nationalist elements in the military in each country was seen as prejudicial to the prospects for success of new liberal projects. Third, the emphasis placed by the United States government and by international institutions in the 1990s on the importance of good governance suggested that the return of the military would have sharply negative international consequences. The new democratic regimes in the subregion sought to engage actively in a range of diplomatic activities and international associations in order to minimize the threat that the military would seek to return to power, and as noted above, MERCOSUR was envisaged, especially in the early stages, as one defence among others against this unwelcome prospect.

This was particularly the case in Argentina, where successive military revolts in January and December 1988 reflected discontent in the armed forces as a consequence of the punishment of members of the former military government for their complicity in the abuse of human rights, and of members of the High Command which had launched the Malvinas invasion. Opposition to the civilian government was particularly intense among the veterans of the Malvinas campaign, and civil–military relations were particularly tense when Menem came to power. Matters came to a head when supporters of former rebel Colonel Mohammed Ali Seneildín seized their military headquarters in December 1990. Menem responded by confronting the rebels, who were eventually dismissed from the service and imprisoned. At the same time, the leaders of the military junta imprisoned for human rights offences were pardoned, and subsequent legislation signalled the end of prosecutions arising out of the 'Dirty War'. By the mid-1990s the military threat had receded, but it was still a live issue as entry to MERCOSUR was negotiated.

THE ENTRY OF PARAGUAY AND URUGUAY

As noted at the outset, the economies of Paraguay and Uruguay are tiny in comparison with those of Argentina and Brazil. They may be regarded as followers rather than leaders in the integration process in the subregion represented by MERCOSUR, but their entry may still be understood in terms of the framework laid out above – the clustering of economic, political, ideological and security concerns in precarious domestic projects focused on achieving and consolidating a transition to liberal democracy, and promoting neo-liberal economic reform.

In the case of Paraguay, political considerations may be regarded as paramount in the motivation of the regime, although the extent of penetration of the economy by Brazilian interests in particular would have made a decision to remain outside an integrated common market in the subregion difficult to contemplate. Paraguay remained under the Stroessner dictatorship (a hybrid regime backed throughout by the Colorado Party, and maintained by its monopoly on state resources) from 1954 through to February 1989, when it was overthrown by General Andres Rodriguez. His subsequent election to the presidency on 1 May 1989 as the candidate of the Colorado Party was more a renewal of the old regime than the creation of a new one. The Rodriguez administration followed a partially neo-liberal economic programme, but sought at the same time to protect the status of the Colorado Party as an unchallenged party of government by strengthening its links with the armed forces and public administration. One consequence of this, reflecting the use of state assets to buy political support for the party, was the reluctance of the newly formed Privatization Council to pass key state holdings into the private sector. His project, in other words, was partial economic liberalism sustained by a privileged party of government.

With 40 per cent of its trade with Argentina and Brazil, its modern growth pole dependent on massive joint hydroelectric projects with the two countries, and its exports dominated by agricultural products (soya and cotton together accounting for 70 per cent), Paraguay had little option but to enter MERCOSUR in 1991. However, a project combining commitment to full democratization with a coherent neo-liberal programme emerged only after entry to MERCOSUR. In the run-up to the 1993 elections Rodriguez was persuaded to announce that he would not run, and the Colorado candidacy was awarded (in questionable circumstances) to entrepreneur Juan Carlos Wasmosy, a business leader made rich from the hydroelectric projects of the 1970s, and favoured by the military establishment. He subsequently won the presidency, but with a minority 40 per cent of the vote.

Two developments in Wasmosy's presidency illustrate the impact of regional politics on the evolving domestic situation. The first is the adoption of a more coherent neo-liberal economic policy. The second, and more important, is the role of regional support for democracy. Initially sponsored by the military, and particularly by First Army corps commander General Lino Oviedo, Wasmosy was seeking by 1994 to impose presidential authority over the hitherto all-powerful army command. At a moment of acute tension at the end of the year, when it appeared that Oviedo might be tempted to seize power, a statement from Brazil's president Fernando Henrique Cardoso spelled out the consequences of such a move – isolation, and expulsion from MERCOSUR. General Oviedo led a rebellion against the regime in April 1996, and was subsequently retired from the Armed Forces. With the loyalty of the new military hierarchy to the elected regime still uncertain, Oviedo continued his opposition to the regime in retirement, eventually winning nomination as presidential candidate for the ruling Colorado while a warrant was out for his arrest. At the time of writing, Wasmosy was threatening to postpone the forthcoming elections (due in May 1998), and MERCOSUR put out a formal statement reminding him that membership required a commitment to the maintenance of national democratic institutions.[24]

The case of Uruguay is slightly different, and reflects the more comprehensive character of the MERCOSUR project as it took shape during the 1990s. Uruguay had returned to democracy under President Julio Sanguinetti in 1985 (elected with 39 per cent of the vote), but relations with the military were not resolved until an amnesty law first enacted in 1986 was confirmed by a referendum in 1989. Sanguinetti's failure to win support for his programme of economic liberalization led to his defeat in 1989 by Lacalle, who also came to power with a minority vote (37 per cent). He in turn sought opposition support through power-sharing in order to press ahead with policies of privatization and liberalization, but had to suffer a modification of the policy by way of a national referendum in 1992. With the return of Sanguinetti to power in 1995 it appeared that democracy was attaining a degree of solidity, although tensions continued over the government's refusal to open up the issue of human rights abuses under the military, and entry to MERCOSUR promised a consolidation of democracy and of precarious economic reforms. At the same time, the opportunity for a role of diplomatic leadership (a natural internal solution to the rival claims of Brazil and Argentina within the association) allowed Uruguay the opportunity to project itself on the international stage. In the Uruguayan case, then, the political, economic, ideological, security and regional dimensions of MERCOSUR were positive and mutually self-reinforcing.

CONCLUSION: THE EMERGENCE OF A REGIONAL ROLE

By the time MERCOSUR came into being on 1 January 1995 there were clear signs that the political and economic situation in the subregion was undergoing a profound transformation. Fernando Henrique Cardoso, the architect of the *real* plan, had been elected to the presidency of Brazil in 1994, and the re-election of Menem in Argentina in May 1995 ensured continuity in the process of implementation of neo-liberal projects in each case. However, it was not yet clear in any of the four member countries that the domestic projects of political and economic restructuring were secure. Despite this, a new process began to develop from this point onwards in which domestic, subregional and regional dynamics converged to strengthen the underlying domestic projects and to give the association a prominent role in regional and potentially in international affairs.

As noted above, the commitment of Argentina and Brazil to economic liberalization, was an essential condition, in view of the weight of their economies, for any meaningful movement towards a free-trade area in the larger region. And it followed that if they were able to combine in an association committed to such a movement, it would have the potential to play a significant role. Through the first three years of its existence it increasingly seemed that the presence and orientation of MERCOSUR was indeed engineering a shift in the balance of power in the region, as rival strategies emerged for the eventual creation of the desired Free Trade Area of the Americas. As a US-inspired vision of the successive incorporation of South American states into NAFTA foundered, as a consequence of suspicion in the US Congress and in South America itself, MERCOSUR sponsored an alternative project in which moves towards a free-trade area would build on NAFTA and MERCOSUR itself without subsuming either. The successive entry of Chile and Bolivia into MERCOSUR as associate members in 1996, in conjunction with the breakdown of negotiations between NAFTA and Chile over Chilean entry, suggested that the 'building block' approach promoted by MERCOSUR would prevail, and allowed President Cardoso to assert that 'the success of the FTAA depends upon the consolidation of MERCOSUR'.[25] Such a development does not constitute a challenge to the strategy of steadily expanded open regionalism, liberalization and privatization, but it does create space for the emergence within that broader project of an alternative perspective to that adopted by the United States. First, MERCOSUR espouses a different order of priorities, which places access to the US market ahead of commitments in the area of intellectual property rights. Secondly, and perhaps of greater consequence, MERCOSUR and its individual members have begun to emphasize the opportunities provided by closer association not with the United States, but with the European Union, where some of its major markets are found. The announcement in February 1998 that the first Latin American–European summit would take place in Rio de Janeiro in 1999 under the auspices of the Rio Group (the international grouping of Latin American states from which the United States is excluded), along with growing interest in Latin American–European trade negotiations on both sides of the Atlantic, suggested that in the longer term MERCOSUR might contribute to a process of economic integration going well beyond the Americas. At this point, then, it is possible to say that the consolidation of domestic projects of democratization and neo-liberal reform in Argentina and Brazil in particular, in part through MERCOSUR, has in turn laid the foundations for the emergence of the subregional association as a significant player in regional and potentially in international politics.

NOTES

1. See J. Grugel, 'Latin America and the Remaking of the Americas', in A. Gamble and A. Payne (eds), *Regionalism and World Order* (London: Macmillan, 1996), esp. pp. 151–3.
2. Data from V. Bulmer-Thomas, 'Regional Integration in Latin America since 1985: Open Regionalism and Globalisation', in A.M. El-Agraa (ed.), *Economic Integration Worldwide* (London: Macmillan, 1997), Table 10.1, p. 256.
3. J. Grugel, 'Latin America and the Remaking of the Americas', in A. Gamble and A. Payne, op. cit.
4. See A. Payne, 'The United States and its Enterprise for the Americas', and J. Grugel, 'Latin America and the Remaking of the Americas', in A. Gamble and A. Payne, op. cit.
5. A. Payne, 'The United States and its Enterprise for the Americas', p. 106.
6. A. Payne, 'The United States and its Enterprise for the Americas', pp. 106-7.
7. J. Grugel, 'Latin America and the Remaking of the Americas', p. 137.
8. United Nations Economic Commission for Latin America and the Caribbean, *Open Regionalism in Latin America and the Caribbean: Economic Integration as a Contribution to Changing Production Patterns with Social Equity* (Santiago, Chile: ECLAC, 1994), pp. 9–16.
9. See W. Robinson, *Promoting Polyarchy: Globalization, US Intervention and Hegemony* (Cambridge: Cambridge University Press, 1996).
10. T.R. Scully, 'Chile: The Political Underpinnings of Economic Liberalization', in J. Domínguez and A.F. Lowenthal (eds), *Constructing Democratic Governance: South America in the 1990s* (Baltimore: Johns Hopkins University Press, 1996), p. 99.
11. See L. de Riz, 'Argentina: Democracy in Turmoil', and B. Lamounier, 'Brazil: the Hyperactive Paralysis Syndrome', in J. Domínguez and A.F. Lowenthal

(eds), *Constructing Democratic Governance: South America in the 1990s* (Baltimore: Johns Hopkins University Press, 1996). The phrase quoted is from pp. 170–71.
12. F.H. Cardoso and E. Faletto, *Dependency and Development in Latin America* (Berkeley: University of California Press, 1979).
13. A. Fishlow, 'A Tale of Two Presidents: the Political Economy of Crisis Management', in A. Stepan (ed.), *Democratizing Brazil* (New York: Oxford University Press, 1989).
14. See W. Smith, 'State, Market, and Neoliberalism in Post-Transition Argentina: the Menem Experiment', *Journal of Interamerican Studies and World Affairs*, 33, 4, 1991.
15. See D.G. Richards, 'Regional Integration and Class Conflict: MERCOSUR and the Argentine Labour Movement', *Capital & Class*, 57, Autumn 1995, p. 68.
16. See D. James, *Resistance and Integration: Peronism and the Argentine Working Class, 1946–1976* (Cambridge: Cambridge University Press, 1988), and M. Keck, *The Workers' Party and Democratization in Brazil* (New Haven: Yale University Press, 1992).
17. D. Rock, *Argentina 1516–1987* (London: I.B. Tauris, 1986), Ch. 6.
18. T. Skidmore, *Politics in Brazil, 1930–1964* (New York: Oxford University Press, 1964).
19. See H. Spalding Jr, *Organized Labor in Latin America* (New York: New York University Press, 1977), Ch. 4.
20. T. Skidmore, *The Politics of Military Rule in Brazil, 1964–1985* (New York: Oxford University Press, 1988).
21. P. Cammack, 'Brazil: The Long March to the New Republic', *New Left Review*, 190, November–December 1991.
22. K. Weyland, 1993. 'The Rise and Fall of President Collor and its Impact on Brazilian Democracy', *Journal of Interamerican Studies and World Affairs*, 35, 1, 1–37.
23. For Argentina, see D. Pion-Berlin, 'Between Confrontation and Accommodation: Military and Government Policy in Democratic Argentina', *Journal of Latin American Studies*, 23, 1991; for Brazil, see W. Hunter, 'Politicians against Soldiers: Contesting the Military in Postauthoritarian Brazil', *Comparative Politics*, 27, 4, 1995.
24. See *Latin American Weekly Report*, WR-97-49 (9 December 1997), WR-98-01 (6 January 1998), WR-98-02 (13 January 1998), and WR-98-09 (3 March 1998).
25. *Latin American Weekly Report*, WR-97-18 (6 May 1997).

6 The Association of Caribbean States

Anthony Payne

The Caribbean has found the deep-seated changes taking place in the structure of the contemporary world order particularly hard to handle. In essence, it fears that the rigours of globalization will expose it as unable to compete effectively in world markets and that the trend towards triadic regionalism will ultimately leave it excluded from the key such organization in the Americas. In short, it worries that it will face marginalization in the new world order. In these threatening circumstances the various states of the Caribbean have sought a measure of comfort in union and have lately formed an Association of Caribbean States (ACS). The ACS is, in fact, an interesting and illustrative example of the contemporary trend towards subregionalism, precisely because it seeks to build linkages between a set of states and societies which have not previously been the subject of serious attempts to effect closer union. It was only formed in July 1994 and is made up of the following 25 states – Antigua and Barbuda, the Bahamas, Barbados, Belize, Colombia, Costa Rica, Cuba, Dominica, the Dominican Republic, El Salvador, Grenada, Guatemala, Guyana, Haiti, Honduras, Jamaica, Mexico, Nicaragua, Panama, St Kitts and Nevis, St Lucia, St Vincent and the Grenadines, Suriname, Trinidad and Tobago and Venezuela. By any standards, this is a large number of countries to seek to bring into regional collaboration and the task is manifestly made the harder by the enormous diversities of situation represented in the membership of the association, ranging from tiny island-states with minuscule populations like Antigua, Dominica and St Kitts to medium-sized states with large populations like Colombia, Mexico and Venezuela (see Table 6.1). Nevertheless, this very diversity is revealing and important in itself, for it draws attention to the fact that the ACS derives its main impetus from a broader political and economic process which is beginning to constitute a reconfiguration of what is meant by, and included within, the concept of the Caribbean.

Table 6.1 Basic Data on Potential Members of the Association of Caribbean States

Country	Area (sq. km)	Population	GDP per capita 1991 ($)
Anguilla	91	8,475	5,929
Antigua and Barbuda	440	79,000	4,969
Aruba	193	65,796	13,475
Bahamas	13,864	254,685	8,205
Barbados	430	257,000	5,750
Belize	22,963	220,000	2,804
Bermuda	53	59,588	23,000
British Virgin Islands	153	14,898	10,538
Cayman Islands	260	28,000	26,160
Colombia	1,141,748	32,980,000	1,254
Costa Rica	50,900	3,032,794	1,871
Cuba	110,860	10,576,921	2,500
Dominica	751	71,183	659
Dominican Republic	48,730	7,320,096	2,148
El Salvador	21,041	5,330,000	1,032
Grenada	344	91,000	1,431
Guadeloupe	1,705	342,175	3,480
Guatemala	108,890	9,467,029	850
Guyana	214,970	710,000	285
Guyane	99,909	114,900	2,180
Haiti	27,749	6,500,000	274
Honduras	112,088	4,800,000	1,040
Jamaica	10,992	2,414,100	1,150
Martinique	1,110	340,381	4,406
Mexico	1,973,000	81,141,000	2,936
Montserrat	104	12,467	7,407
Netherlands Antilles	766	191,674	6,110
Nicaragua	139,000	3,900,000	284
Panama	78,200	2,329,329	2,070
Puerto Rico	9,104	3,300,000	7,755
St Kitts-Nevis	269	45,000	2,979
St Lucia	616	151,290	673
St Vincent	389	107	1,564
Suriname	163,820	401,000	3,501
Trinidad & Tobago	5,128	1,234,388	4,127
Turks & Caicos	430	12,350	6,000
US Virgin Islands	352	110,000	8,717
Venezuela	912,050	19,760,000	2,213

Source: Adapted from *CARICOM Perspective*, July–December 1993, p. 9.

Conventionally, the Caribbean has been defined by a blend of its geography and history. From a simple geographical viewpoint, it consists of all of the islands in the Caribbean Sea, which make up a huge archipelago running some 2500 miles from the southern tip of Florida in the north to the coast of Venezuela in the south. From a more complex historical viewpoint which emphasizes the common impact of slave-based European imperialism, it also includes Belize on the Central American isthmus and the three 'Guianas' on the South American coast: Guyana, Suriname and French Guiana. Conceived in this way, the region has widely been said to possess an intellectual coherence that makes it possible to analyse its modern politics and economics within a single framework.[1] Yet, by simply extending the geographical approach to include all those territories whose shores are washed by the Caribbean Sea, another plausible definition of the Caribbean is advanced. By this measure, the region would also include Mexico (or at least the Yucatan province of Mexico) and the Central American states of Honduras, Guatemala, Nicaragua, Costa Rica and Panama (with El Salvador generally added despite the fact that it has a Pacific, rather than Caribbean, shoreline because it is part of Central America and cannot, as it were, be left out of groupings which embrace all of the other states of the isthmus), as well as Colombia and Venezuela. The problem with this latter notion of a Caribbean Basin, as we shall see, is that it lacks historical credence. It is not based on a long-standing internalization of the concept within the thinking of political leaders or intellectuals in either the Caribbean or Central America[2] and can thus be said to lump together the two main parts of the putative Basin in a largely artificial way.

This chapter therefore explores how and why it was that the ACS came to be formed in the mid-1990s as the main institutional vehicle by means of which an attempt was to be made to give the concept of a Caribbean Basin meaningful economic and political shape. It starts by identifying the major historical legacies which impinge on region-building within the Basin and, in so doing, inevitably identifies divisions relating to matters of culture and ideology; goes on to look at the series of diplomatic steps which led to the establishment of the ACS; sets out the basic framework and describes the early activities of the new association; and ends by trying to assess some of the problems which the ACS is currently facing and which are likely to imperil any subregionalist experiment in the contemporary Caribbean Basin. In this last section some wider connections are also made to the plight of these small states in the world order at the end of the twentieth century and reference is made in particular to some of the new political economy and new security dilemmas which such states now face.

BACKGROUND CONSIDERATIONS

As already suggested, the history of the Caribbean Basin is an unpropitious basis on which to seek to build effective subregionalism. Given the constraints of space, four points must suffice to make this argument.
1. The Basin is divided in the most complex way between the historical legacies of different European colonialisms. The major divide is between Hispanic and Anglophone cultural, linguistic and political traditions. This immediately separates out the English-speaking territories of the Caribbean, with their traditions of parliamentary government, inheritance of an English legal framework and long-standing lines of communication with London. A genuine West Indian identity does bind these countries together, although it has always been a weaker force politically than narrow territorial insularism.[3] There exists no such underlying Hispanic unity to the rest of the Basin, other than in the broadest cultural terms. The Central American countries have never had their entire societies reconstructed according to the demands of imperialism, the fate which befell the Hispanic Caribbean, even though, like the Dominican Republic and Cuba, they too suffered the consequences of Spanish rule. Mexico has historically looked more to North America, and Colombia and Venezuela more to South America; their greater size has also tended to make them stay aloof from other Caribbean Basin states. Additional linguistic and cultural complications derive from both Francophone and Dutch cultural and political influences in other parts of the Basin. Indeed, to this day some Caribbean territories are incorporated into the metropolitan frameworks of France and the Netherlands. Britain also has continuing dependencies in the region, which means that difference of constitutional status acts as a further dividing factor.[4]
2. By contrast, the impact of European colonialism has been broadly similar in economic terms across the Basin. As Malcolm Cross put it several years ago, all the countries of the Basin have either 'had to, or still have to, come to terms with the uniquely New World experience of being dependent suppliers of tropical primary products for Western European or North American markets'.[5] This is not to imply that important economic differences do not presently exist between Basin states. As has been noted already, some are obviously larger than others in area and population and are better endowed with resources, especially perhaps those with oil. Some states have advanced manufacturing sectors; others are making their initial moves in this direction; still others remain reliant on agricultural commodity production as their main economic activity. Yet one of the real

consequences of the kind of broad historical unity in terms of political economy which can be said to attach to the Basin is that many Basin states are now rivals with each other in key sectors of their national economies. This applies whether they are competing to sell bananas, attract tourists or infiltrate nascent manufactured goods into the North American market. In principle, there is a basis for economic cooperation across the Basin; in practice, at least as it has unfolded thus far, competition has been the more striking characteristic. This was apparent even in the era of import-substitution industrialization in the 1950s and 1960s and inevitably became even more marked a feature as virtually all Basin states moved to espouse liberal, export-oriented strategies of growth during the 1980s.[6]

3. The Basin has been the scene already of a number of attempted regional integration movements. These represent what, in the terminology of this book, would have to be called 'sub-subregionalisms'; one even constitutes 'sub-sub-subregionalism'. Thus the English-speaking Caribbean states have been gathered together within the Caribbean Community (CARICOM) since 1973 and the Caribbean Free Trade Area (CARIFTA)[7] before that; the smallest eastern Caribbean islands constitute the Organization of Eastern Caribbean States (OECS)[8] as a sub-set of CARICOM; the Central American Common Market (CACM)[9] has linked the isthmus states economically since 1960; and Mexico, Colombia and Venezuela now constitute the so-called Group of 3 (G3). In other words, all past attempts at building regionalism in the Basin have attempted to do this by defining their notion of 'region' in more specific and more limited ways. These various groupings all still exist in varying stages of health and development and represent political interests which have to be brought on board within any Basin-wide process of collaboration.

4. Within the Caribbean Basin the notion of the Basin carries with it an unfortunate sense of external manipulation, to the point where some would say that it was an ideological construct that only existed in the minds of those outside the Basin, particularly within the United States. Much of this unease was generated in the 1980s during the Reagan presidency when the phrase captured the boundaries of a region deemed ripe for revolution and in which fundamental US security interests were considered to be at stake. Traditional distinctions between the policy approaches adopted towards the Caribbean and Central America were abandoned and US policy was indeed a Basin-wide policy for the duration of this period, as witnessed symbolically by the administration's promotion of the Caribbean Basin Initiative (CBI), the trade and aid package proposed to the US Congress by Reagan in 1982. Different parts of the Basin were positioned differently within this programme, and Cuba, Nicaragua and Grenada

were entirely excluded.[10] Nevertheless, the explicit adoption of a Basin framework by the US required some measure of collective response by recipient states and led, for example, to such innovations as the formation of an informal CBI group of ambassadors in Washington. The negative consequence from a contemporary subregionalist perspective is that the Basin concept is now unavoidably tainted to some degree by past US geopolitical projections.

There is much that has necessarily been glossed over in this brief account of the historical backcloth to the creation of the ACS. Nevertheless, the point to emphasize is that there was very little in the inherited historical structures of the region on which an emergent Caribbean Basin subregionalism could expect to feed. It was never likely to be an organic development. The reality instead was that a good deal of history would have to be overcome if such a movement were ever to prosper. This alone suggests that exogenous forces may have been decisive in launching the process which led to the ACS's formation and that, in particular, the perceived need to react collectively in some way to US prognoses for the subregion may have been critical.

ORIGINS

The best place to look for the origins of the new Caribbean Basin subregionalism is, perhaps predictably, in the failings of the old sub-subregionalism, in particular the difficulties into which CARICOM was evidently running by the end of the 1980s. From its inception CARICOM had confronted a dichotomy which came to be referred to within the organization as 'deepening' versus 'widening'. The former argument highlighted the case for advancing the intensity of cooperation on a range of fronts between the core Anglophone Caribbean membership of CARICOM; the latter advocated a minimalist programme based around free trade and functional linkages but extended to include the wider Caribbean. (It is interesting to note that within the Anglophone Caribbean, expansion to Central America was never considered, even by the most ambitious proponents of 'widening', which is itself revealing of the substantial cultural gulf and consequent lack of practical contact between these two leading component sections of the Basin). Within CARICOM the argument against 'widening' that was always made was that it would dilute Anglophone control of the regional integration process, since nearly all the actual or potential candidates to join (Haiti, the Dominican Republic and, more recently, Venezuela) were both demographically much larger than, and culturally and linguistically distinct from, the founding members. The risks of such 'widening' were only deemed acceptable if sufficient 'deepening' had already taken place amongst the core. In the event, neither was achieved. CARICOM survived as an institutional apparatus, but at the price of stagnation.[11]

Over time, though, its failures were increasingly recognized and in 1989 a major enquiry was launched into its functioning by an independent West Indian Commission, a group of distinguished figures from the Anglophone Caribbean led by Sir Shridath Ramphal, the Guyanese former Commonwealth Secretary-General. What eventually prompted the decision of the various CARICOM heads of government to embark upon such an extensive review was their realization, as they put it, of 'the challenges and opportunities presented by the changes in the global economy'.[12] Chief amongst these from their point of view at that time was the approaching creation of a single market within the European Community (EC) by the end of 1992. This threatened the traditional preferential access historically enjoyed by the English-speaking Caribbean's agricultural products, especially bananas, to their main (and former imperial) overseas market. More broadly, it also seemed to be characteristic of the emergence of new regionalist centres of power within the global political economy which looked as if they might damage CARICOM. Such an anxiety was given added force during the course of the West Indian Commission's deliberations when President Bush's Enterprise for the Americas Initiative (EAI), announced in June 1990, immediately set in train the prospect of free trade throughout the Western hemisphere, pioneered by the likely negotiation of a North American Free Trade Agreement (NAFTA) linking the US with Canada and Mexico. The Commission sought ultimately to resolve the impasse facing CARICOM by proposing in its final report a simultaneous 'deepening' and 'widening' of the Community. However, its key proposal in the first regard – the establishment of a permanent Caribbean Commission of three former political leaders designed to drive forward the internal integration process – was rejected by the heads of government. They acknowledged the problem that CARICOM needed a stronger executive agency to see that decisions taken at summit meetings were indeed subsequently implemented, but preferred to fall back upon a feeble compromise whereby a so-called CARICOM Bureau, composed of the past, present and next chairs of the heads-of-government summit, be set up and charged with filling the vacuum that had been detected. As a consequence, CARICOM of itself did not take the big step forward demanded by the West Indian Commission.

However, the Commission's solution to the 'widening' dilemma aroused interest and did generate support. This was the proposed creation of a new body, a putative Association of Caribbean States, anchored on CARICOM but open to the wider Basin region. This reflected a new perspective within the CARICOM subregion, a shift towards a pro-active approach towards cooperation with their non-Anglophone neighbours, based in good part on a growing awareness that a grouping of only some 5.5 million people was simply too small to meet the challenges of a fast-changing global political economy. As the West Indian Commission itself argued in its report:

> Put simply, the peoples of CARICOM and their Governments must no longer think in narrow terms merely of 'the Commonwealth Caribbean', but in wider terms of a 'Caribbean Commonwealth' – and must work to fulfil this larger ambition. The ambition itself must encompass, besides the 13 CARICOM Countries, all the countries of the Caribbean Basin. It must reach out, therefore, to all the independent island states of the Caribbean Sea and the Latin American Countries of Central and South America whose shores are washed by it. And it must be open as well to the Commonwealth of Puerto Rico, the island communities of the French West Indies, the Dutch islands of Aruba and the Netherland Antilles, the US Virgin Islands and the remaining British dependencies.[13]

The ACS was envisaged as 'being functionally active in an integration sense'[14] and a wide set of possible areas of cooperation was listed, including the negotiation of special trading terms, the widening of communication links, cooperation in tourism and health matters, the management of the resources of the Caribbean Sea and the curbing of drug trafficking. Nevertheless, the aim was quite deliberately not to be too specific at too early a stage – for it was always recognized that formidable constitutional and cultural difficulties, as identified in the previous section of this chapter, lay in the way of the project – but rather to give voice to the idea and subsequently to raise it with the governments of other Basin countries.

The ACS idea was accepted in exactly these terms by the CARICOM heads-of-government conference which met in October 1992 to consider all the recommendations of the West Indian Commission. The new CARICOM Bureau subsequently suggested that an initial focus on 'a selected range of issues' would 'help to define the association' and identified intra-group trade liberalization, functional cooperation and the development of a group relationship with the rest of the hemisphere as the immediate priorities.[15] Michael Manley, the former prime minister of Jamaica, was thereafter asked out of retirement, as it were, to conduct what was called a 'probe' of other non-CARICOM views of the proposal. As a result of his mission in the summer of 1993, a group of technical staff was gathered together and charged with preparing a paper on an organizational structure for the Association in sufficient time for it to be possible to sign a document by mid-1994. There was for a short period a danger that the whole project would be seen in Central America and the other non-English-speaking parts of the region as too much of an extension of CARICOM. The tenor of the discussion of the proposed ACS at the initial CARICOM heads-of-government meeting had by all accounts seemed to presume that the Anglophone Caribbean countries would be unquestionably at the core of such a wider grouping, notwithstanding the fact that, as we have seen, CARICOM represents only a small part of the wider Basin region. However, CARICOM's leaders were alerted to the risk of giving this impression and the governments of Colombia, Venezuela, Mexico and Suriname remained fully part of the planning process.

Indeed, the leaders of these four countries joined with CARICOM leaders in a unique meeting in Port of Spain, Trinidad in October 1993 which not only expressed public support for the ACS idea but ended with the signing of an action plan on interim economic cooperation. This ranged over matters such as trade, investment, tourism and transport and was limited in its specific commitments. Its longer-term significance lay in the fact that all the signatories were committed to a model of economic development which gave priority to openness rather than protection. It is important to note that at no stage did ideological divisions over core questions of political economy threaten the building of the ACS. CARICOM's leaders thus particularly welcomed the recent decision of the G3 countries – Colombia, Mexico and Venezuela – to establish a free-trade arrangement amongst themselves to coincide with the inauguration of the NAFTA, which by then had been negotiated successfully and was due to come into effect on 1 January 1994. Venezuela had concluded a one-way free-trade deal with CARICOM a year earlier and one outcome of the Trinidad meeting was the likelihood of a similar agreement being signed between CARICOM and Colombia by the end of 1994. For his part, President Salinas de Gortari of Mexico formally welcomed the possibility of free trade with CARICOM and it was announced that discussions along these lines would also be initiated with Mexico. Although deciding to proceed on a bilateral basis, the idea of joint free trade involving CARICOM, Suriname and the G3 was clearly at the forefront of the discussion. Mexico's position was the critical one, since its involvement would have served in effect to attach the whole grouping to NAFTA. As for Suriname, it was now considered the only other independent Caribbean state eligible for actual membership of CARICOM (and indeed its admission as the fourteenth and only non-Anglophone Caribbean member was approved by the CARICOM heads of government in February 1995).

All this early diplomacy was further complicated by the positions of the other important Caribbean states that were not members of CARICOM. The traditional strategy of the Dominican Republic, for example, has been to straddle the Caribbean and Latin America, participating in the Lomé Convention with the EC along with the CARICOM countries, even at one time in 1989 seeking membership in CARICOM, and yet, in other respects, working more closely with its Hispanic rather than its Anglophone Caribbean neighbours. Its foreign policy on regional matters has thus rarely been grounded firmly. However, the Dominican Republic indicated at a fairly early stage in the debate that, in principle, it too would wish to join the ACS, and a government commission, headed by Vice-President Carlos Morales Troncoso, was established to study the merits of closer involvement with Caribbean regional integration projects. For its part, Cuba was also keen to offset its general post-Cold War isolation by developing any kind of contact with its regional neighbours. The Cuban government accordingly pressed hard to get the ACS process moving and its foreign minister, Roberto Raina, conducted an extensive tour of Venezuela and no less than seven CARICOM countries in November 1993, talking the language of economic and cultural cooperation. CARICOM's leaders had in any case agreed earlier in the year to seek to promote ties by means of the establishment of a CARICOM–Cuba Mixed Commission similar in concept to existing joint bodies with Mexico and Colombia. Some also expressed their support for Cuba's admission into the ACS as a founding member. However, the idea of the Mixed Commission had not been well received in the US and the continuing obduracy of the US position on all matters relating to Cuba unquestionably acted as a brake on the island's re-embrace within Caribbean organizations. By contrast, the military regime in Haiti manifested no interest in joining the ACS, aware no doubt that its overthrow of the democratically elected government of Jean-Bertrand Aristide in September 1991 made it politically unacceptable to the rest of the subregion. On the other hand, the exiled Aristide, who received strong diplomatic support from other Caribbean Basin governments, particularly Venezuela, was often represented at the various ACS preparatory meetings.

Nevertheless, although the positive reactions of Santo Domingo and Havana were important and encouraging developments, the real prospects of the ACS always depended on the reaction of the Central American states. Prior even to the publication of the West Indian Commission Report, there had been some moves to forge closer Central American and Caribbean links. This initial impetus undoubtedly derived from consideration in the region of what was the most appropriate negotiating unit with which to relate to the EAI and the NAFTA initiatives emanating from the Bush administration in Washington. It is significant too that one notes here for the first time the role of another actor in the integration process besides politicians and technocrats. Encouraged by the Washington-based business pressure group Caribbean/Latin American Action (C/LAA), the private sector of the Basin took a lead and began to argue the case for greater subregional cooperation in response. The government of Governor Rafael Hernandez Colon in Puerto Rico (which is constitutionally a self-governing 'Commonwealth' associated with the United States) also sought to promote closer links between Central American and Caribbean political leaders, inviting both Honduran President Rafael Callejas and Michael Manley, who was still then the Jamaican prime minister, to the regular Caribbean Basin business conference in San Juan in August 1991 and setting up the so-called Caribbean Basin Technical Advisory Group (CBTAG) to facilitate joint policy-making between the two parts of the Basin. Callejas and Manley got on well together, met again at the C/LAA Miami Conference in December and set in train the first ever joint meeting of Central American and Caribbean foreign ministers which took place in San Pedro Sula, Honduras in January 1992. CBTAG itself constituted, albeit in embryo, a workable institutional framework for closer Caribbean Basin cooperation and potentially was capable not only of organizing relations with the EAI and the NAFTA but also of promoting regional economic integration within the Basin itself.[16] Bringing together as it did public-sector technocrats, private-sector business people and academics from across the whole of the subregion, it was undoubtedly a highly creative innovation by the Puerto Rican government and a genuinely novel moment in the affairs of the countries of the Caribbean and Central America.

However, despite this promising beginning the new Basin-wide dialogue swiftly ran into difficulties. The Puerto Rican administration which had sponsored CBTAG was defeated in elections in November 1992 and replaced by a regime unconcerned with the wider Basin and committed as its top priority to the negotiation of statehood for Puerto Rico *within* the US. As a result, CBTAG was, to all intents and purposes, disbanded and Puerto Rico substantially withdrew from the regional diplomatic scene. As already indicated, the early discussions of the ACS also appeared to be too CARICOM-orientated from a Central American perspective. Most importantly, both reflecting and exacerbating these problems, there developed during 1992 a major row over bananas. The context was the attempt by the EC to reconcile its promise under the Lomé Convention to provide a preferential market in bananas for producers within its former colonies with its commitment to establish a single European market. The EC's dilemma was the more acute still because the case against the continuing protection of bananas was also strongly being made by the US and Latin America within the Uruguay Round of the GATT. As regards the Caribbean Basin, it placed different parts of the Basin in competiton with each other in the most direct way. On one side were the smaller Caribbean banana-producing states, mainly in the eastern Caribbean but also including Jamaica, unable for complex historical and cultural reasons to grow and ship bananas cheaply enough to sell in an open world market; on the other were the more commercially competitive so-called 'dollar-banana' states of Central America, where banana plantations were typically owned and/or controlled by large US-based corporations which chafed at the constraints placed by the EC on their access to the European market. An eventual, highly complicated compromise, based on tariffication rather than quotas, was only struck in mid-December 1992 after much lobbying and political conflict. It was not as good a deal for the English-speaking Caribbean as it had previously enjoyed under Lomé, but it was still too protectionist from the point of view of Central American producers.[17]

All this, to state the obvious, scarcely helped engender friendly relations on other important regional issues in the emerging Caribbean–Central American dialogue. It seemed for a while as if the political discord arising from the banana dispute would linger on, with the Central American leadership apparently backing efforts to challenge the new regime in the European courts. However, the secretariats of CARICOM and the CACM met in Guatemala City in February 1993 and a second joint meeting of foreign ministers did take place in Kingston in May 1993, with Manley's 'probe' of key Central American countries in connection with the ACS helping to reduce tensions still further. The intention of both sides appeared to be to set the banana issue aside as much as possible in order to facilitate collaboration on other matters, with the result that the Central American countries publicly voiced their support for the ACS at the end of the second foreign ministers' conference. Manifestly, though, they did not defer to Caribbean leadership on the

ACS; some were attracted by the potential of the work of CBTAG and still saw the case for wider Basin cooperation in negotiation with the US, especially given that the issue of trying to ensure so-called NAFTA 'parity' for the CBI countries in respect of trade access to the US market, and by extension the bigger question of possible entry into NAFTA itself, were just coming on to the agenda. The same applied potentially to the EC with which the Central American countries had themselves been linked since the mid-1980s under the terms of the San José process. Nevertheless, their dominant view was that the Caribbean was the weaker half of the Basin and needed, as it were, to come to them, rather than the reverse. They were certainly not prepared to tolerate a CARICOM attempt to control the agenda of the ACS and, for a while, it was not at all obvious that they would actually join. Indeed, given the absence of an appropriately supplicant approach (as they saw it) on the part of the CARICOM states, the inclination of many was to seek to reorganize their relations with the US either bilaterally or, at best, on a common Central American basis.

The question of how to establish the right relationship with the emerging NAFTA was thus absolutely central. From March 1993 onwards Caribbean and Central American governments were able to join together to back legislation tabled in the two parts of the US Congress by the representatives of the state of Florida. This putative 'Caribbean Basin Free Trade Agreements Act' would have given the CBI countries parity with Mexico in terms of tariffs, rules of origin and quota elimination for a three-year transitional period during which time they were expected to pursue programmes of liberalization and adjustment. At the end of the transition they were supposed to seek some form of access to the NAFTA on the basis of trade reciprocity. Initially, the new Clinton administration did not express a firm view on these bills, indicating only its willingness to have them submitted to the Office of the Trade Representative for study. In general on this question, the political signals were mixed. However, Clinton eventually was persuaded to meet with the leaders of the four largest CARICOM states – Jamaica, Trinidad and Tobago, Barbados and Guyana – in August 1993 and he promised at the end of these discussions to instruct the US Trade Representative, Micky Kantor, to undertake a study of the impact of the NAFTA on the smaller Caribbean economies. However, no progress was made in the Congress on the 'Caribbean Basin Free Trade Agreements Act'.

The point is that, from the perspective of the CBI countries, which were bound together more effectively by their concern for the future of this programme and the subsequent terms of their access to the US market than any wider sense of Basin unity, the concepts of parity and transition constituted a formula around which everyone could gather for the time being. If passed, the Florida act (or something like it) would obviously have given all parts of the Basin a breathing space. But it did not, and could not, avoid the deeper questions posed by the whole EAI and NAFTA process. The restructuring of the world order generated by the ending of US global hegemony and the demise of the Cold War placed the Caribbean and Central America, for virtually the first time in their history, in a common political and economic predicament. This was the key contextual consideration which kept the various governments of the Caribbean Basin focused on the ACS proposal and, in the end, brought about just enough agreement for a Convention establishing the ACS to be signed at Cartagena in Colombia in July 1994.

ESTABLISHMENT AND FUNCTIONING

The ACS undoubtedly offers a new regional architecture for Caribbean Basin subregionalism. The Convention was signed by all of the independent states initially deemed eligible, with the sole exception of El Salvador, which joined a little later to bring membership up to the intended figure of 25. However, in deference to the complicated constitutional character of the region, the Convention spoke deliberately of bringing together the peoples and economies of all Caribbean Basin 'states, countries and territories'.[18] This wide-ranging formulation allowed several other non-independent countries and territories to be listed as eligible for associate status, namely: Anguilla, Aruba, Bermuda, the British Virgin Islands, the Cayman Islands, Montserrat, Puerto Rico, the Turks and Caicos Islands, the US Virgin Islands, the French *départements* of Guadeloupe, Martinique and Guyane, and the Netherlands Antilles. France signed the Convention on behalf of its Caribbean departments, causing no little anxiety about the role that it might seek to play in the subsequent development of the organization. Puerto Rican representation was notable by its absence, constrained by its association with the US and the latter's disinclination to have any involvement in a body which had admitted Cuba. The degree of involvement of the British and Dutch dependencies remained to be clarified. In other words, in its very composition the ACS unavoidably mirrors the fragmented reality of the contemporary Caribbean political landscape.

The tone of the Convention was also bland, perhaps necessarily so at such an early stage in the development of a new organization. The ACS was defined in Article III.1 as 'an organization for consultation, cooperation, and concerted action'. Its purpose was to identify and promote the implementation of policies and programmes designed to:

1. Harness, utilize, and develop the collective capabilities of the Caribbean region to achieve sustained cultural, economic, social, scientific and technological advancement;
2. develop the potential of the Caribbean Sea through interaction among member states and with third parties;
3. promote an enhanced economic space for trade and investment with opportunities for cooperation and concerted action, in order to increase the benefits which accrue to the peoples of the Caribbean from their resources and assets, including the Caribbean Sea; and
4. establish, consolidate, and augment, as appropriate, institutional structures and cooperative arrangements responsive to the various cultural identities, developmental needs, and normative systems within the region.

Article III. 2 of the Convention then went on to specify that, in pursuit and fulfilment of these purposes, the ACS should gradually promote among its members the following activities:

1. Economic integration, including the liberalization of trade, investment, transportation, and other related areas;
2. discussion on matters of common interest for the purpose of facilitating active and coordinated participation by the region in various multilateral forums;
3. the formulation and implementation of policies and programmes for functional cooperation;
4. the preservation of the environment and conservation of the natural resources of the region and especially of the Caribbean Sea;
5. the strengthening of friendly relationships among the governments and peoples of the Caribbean; and
6. consultation, cooperation, and concerted action in such other areas as may be agreed upon.

As will be obvious from the above, the intended dynamic was that of intergovernmental cooperation in key areas of economic and functional cooperation, including the important matter of coordination of positions in external negotiations. This reflected the initial CARICOM conceptualization of the ACS. Indeed, this traditionalist vision of the nature of the new organization was never seriously challenged at any point in the formative process, perhaps out of a disinclination by the larger states not to be seen to be imposing themselves on their smaller colleagues. Yet, as Henry Gill, the consultant who wrote the early statements of position for the CARICOM secretariat, has admitted, the reality was that 'CARICOM countries had not considered alternative conceptual models, nor had they come to the meetings with fully considered positions on all aspects of their own proposal.'[19] Accordingly, the institutional structure of the ACS is entirely straightforward. Its permanent organs are a ministerial council and a secretariat. The former is to be the principal policy-making body and will hold regular annual meetings and others as necessary. It is empowered by the Convention to establish other committees to assist it in the performance of its functions, with the following specifically named:

- The Committee on Trade Development and External Relations
- The Committee for the Protection and Conservation of the Environment and the Caribbean Sea
- The Committee on Natural Resources
- The Committee on Science, Technology, Health, Education, and Culture
- The Committee on Budget and Administration.

The secretariat was always intended to be small in size, with an explicit commitment to work with other existing regional administrative units. These include not only the secretariats of CARICOM, the OECS and the CACM, but such bodies as the offices of the UN Economic Commission for Latin America and the Caribbean (ECLAC) in Port of Spain and the Latin American Economic System (SELA) in Caracas. Predictably, something of a battle took place over its location. Trinidad and Tobago eventually emerged from CARICOM discussions as the choice of the governments of that particular sub-subregion, Belize and Jamaica having agreed to fall into line behind the candidature of Trinidad, and was the site eventually chosen after consideration had also been given to the cases advanced by the Dominican Republic and Venezuela. The *quid pro quo* was that the first secretary-general of the ACS would be chosen from a Spanish-speaking, rather than English-speaking, member state. This immediately gave an insight into the pattern of bargaining likely to be needed to resolve contentious issues and of the attention that would have to be paid at all times to the mutual sensitivities induced by the cultural chasm which the ACS has necessarily to bridge.

Finally, on organizational matters, it should be noted that the Convention in Article IX(d) did provide for the involvement of 'social partners', defined as non-governmental organizations or other entities representative of wider interests. The inclusion of such a concept actually proved to be an awkward issue during the negotiations, not so much as a matter of principle, but because certain delegates were concerned that 'this could open the floodgates for all types of organizations'.[20] The solution reached allowed for case-by-case recognition of eligibility by the ministerial council and at least created the opportunity for a process of lobbying and the wider participation of social actors within what would otherwise have been a narrow and unnecessarily restricted forum for intergovernmental bargaining. The longer-term development potential of the ACS has thereby been increased as a result and some respect paid to the important role played by regional private sector organizations in building up support for the whole process of Caribbean Basin subregionalism.

It is, of course, very early to offer a view as to the actual functioning of the ACS. On 1 January 1995, the day it was supposed to start its operations, the Trinidad and Tobago government had yet to provide the awaited headquarters. Disputes about the secretariat's budget also took time to resolve, with only the modest sum of US$1.5 million per annum being approved. Signatory states were slow to ratify the Convention and the necessary two-thirds figure needed to bring it legally into force was only achieved a week or so before the inaugural summit meeting was held in Port of Spain in August 1995. Nevertheless, the gathering was a success, with most countries being represented by their president or prime minister. A Venezuelan economist, Dr Simón Molina Duarte, was elected unopposed as secretary-general and Mexico agreed to chair the association for its first year.

More importantly, a declaration of principles and a plan of action were approved. The former expressed the commitment of the leaders to the initiation of 'a new era in the Caribbean region characterized by the strengthening of integration, concerted action and consultation' and reiterated their claim that the ACS provides 'an ideal mechanism' and 'a unique opportunity' for 'responding to the challenges and opportunities presented by the globalization of the world economy, increased trade liberalization and competition for investment and markets', thereby once again drawing attention to the essentially 'reactive' nature of the whole ACS process.[21] The latter identified tourism, trade and transportation as 'three critical sectors' to be accorded immediate priority and highlighted in particular the need to promote trade liberalization, to cooperate more effectively in tourism (which was recognized as the single economic activity common to all members) and to develop better and more integrated regional transport systems. Several Central American delegates in fact observed in the discussions that they had been obliged to travel via Miami to reach Port of Spain. This reflected the cumulative consequences of the historic tendency of transportation systems in the subregion to concentrate provision on connections 'outwards' to the particular imperial centre rather than 'inwards' to immediate neighbours and the new contemporary reality of Miami's emergence as the effective transport hub for the whole of the Basin. Nevertheless, the complaint served effectively to illustrate how many practical and logistical difficulties will be encountered on the road towards efficacious Caribbean Basin subregionalism.

Since the inaugural summit two meetings of the ministerial council have also been held – in Guatemala City in December 1995 and in Havana in December 1996. The first meeting approved the association's initial work programme; the second decided to set up a regional integration fund to finance projects aimed at strengthening trade, tourism and transport links across the ACS. The fund will operate with voluntary contributions from countries, international organizations and public and private entities. Outside the ACS framework, but very relevant to its further development, the CARICOM and Central American states agreed at the beginning of 1997 to start negotiations on a joint free trade agreement.

CONCLUSION

As indicated, the ACS is still a young body and its prospects should not be written off too soon. Yet, manifestly, it already faces several major challenges, all of which revolve around the key question of whether it can really add anything of political and economic significance to the myriad of other sub- and sub-subregional organizations which litter the Caribbean Basin. In the final analysis, what is the ACS for? What is it that this organization can do to help the small states of the Basin find a satisfactory niche in the emerging world order which other existing mechanisms of regional collaboration cannot achieve? What could be said to be its potential *distinctive* contribution to the future development of the Caribbean Basin? Or is it revealed to be an essentially reactive, rather than proactive, response to threatening times?

The reality is that the Basin faces two fundamental challenges at the present time – one, as indicated at the outset, a matter of political economy, the other a matter of security. The key political economy issue confronting all the states of the Basin is their future relationship with the United States, NAFTA and the proposed Free Trade Area of the Americas

(FTAA), agreed at the Miami 'Summit of the Americas' in December 1995 and due to be negotiated and set in place according to the declared wish of the hemisphere's heads of state by the year 2005. Given the combination of the unavoidable constraint represented by small domestic markets and the extent of the commitment to export-led growth presently espoused by virtually all Caribbean Basin governments, the specification of the terms of trading access to the huge US market is the critical matter of contemporary economic policy in the Basin. Yet the precise steps by which an FTAA can be negotiated are not clear to anyone and are inevitably hugely complicated by the prior existence of NAFTA. As regards the Basin, policy on this front is hardly susceptible to ACS coordination since Mexico is both a member of NAFTA and the ACS and has already made it clear that it does not see 'foreign policy' as a priority for ACS activity. In any case, the Mexican government will not want to prejudice the privileged access to the US market which NAFTA already gives it by comparison with all other Latin American and Caribbean states. Basin countries are likely to find that they will have to treat with the US on the FTAA either separately, or in sub-subregional groupings such as CARICOM and CACM, or at best perhaps in a joint CARICOM–CACM negotiation. It is difficult, to say the least, to see how the ACS can become the vehicle by means of which this crucial matter is resolved.

The other key challenge facing the Basin concerns security, in particular the threat to political order in the subregion represented by the rampant drug-trafficking which has grown up so extensively over the last decade. The sheer extent of the financial resources which the drugs cartels now command mean that they can easily subvert the normal processes of administration and justice in any small state where the lawful incomes of politicians and public officials are limited by the facts of historic underdevelopment. Indeed, it is no longer an exaggeration to talk of the possible emergence of 'narco-states' in several different parts of the Basin. By this is meant the effective, but formally unacknowledged, take-over of the governing process of whole states by narcotic interests, a corruption of politics which has perhaps come closest to occurring in Antigua and Colombia. The appropriate security response to this kind of threat is hard to discern, even in principle, let alone in practice, and it is probable that the solution is beyond the internal capacities of Basin states even acting collectively. Yet it is still striking that the ACS has no profile at all in this sphere of policy. The Declaration of Principles adopted at the inaugural summit included a ritual pledge 'to strengthen cooperation within the region to combat the drug menace which seeks constantly to undermine good order and social stability',[22] but there has been no attempt to follow this up with any action. The harsh reality is that the ACS is simply not an organization with any relevance to the new security agenda of the Basin.

In these circumstances the work of the ACS is reduced to the troika of tourism, trade and transportation highlighted in its official plan of action. The chosen focus on these policy areas is entirely sensible in a functional sense but, for complex historical reasons, they will all be difficult matters on which to make visible, politically significant progress. As no less a figure than Fidel Castro told the 1995 summit: 'our common trade is insufficient. Our economies do not complement each other. Traditionally we have conducted our trade with the developed nations and our domestic markets are weak.'[23] Faced with such a situation there is an obvious danger that the ACS will become bogged down in a whole range of awkward constitutional matters, such as the terms of the participation of the non-independent territories and, in particular perhaps, the role of France and its relationship within the context of the ACS to its Caribbean *départements*. If this happens, the major states within the association will not find it an important enough body to which to devote attention and political energy and will start to attend less frequently or at a lower level of representation. The potential benefits of gradual elite socialization into a new Basin-wide way of thinking and cooperating will thus be lost and the many, still existing lines of division within the subregion (of which bananas remain the most potent) will resurface with destructive force. In effect, the opportunity to build an effective Caribbean Basin subregionalism as the platform from which all the various small states of the Basin could manage the pressures of the new global order will have been missed.

NOTES

1. For an elaboration of these considerations, see A.J. Payne, *The International Crisis in the Caribbean* (London: Croom Helm, 1984), pp. 1–10.
2. Dr Eric Williams, the distinguished historian and former prime minister of Trinidad and Tobago for many years, put the point in typically pithy fashion in condemning in the mid-1970s a plan to call a conference on the Caribbean Basin, 'whatever that may be'. *Trinidad Guardian*, 16 May 1975.
3. For interesting discussions by believers in a West Indian identity, see S.S. Ramphal, *West Indian Nationhood: Myth, Mirage or Mandate?* (Georgetown: Ministry of External Affairs, 1971) and W.G. Demas, *West Indian Nationhood and Caribbean Integration* (Bridgetown: Caribbean Council of Churches Publishing House, 1974).
4. These continuing European links are extensively explored in P.K. Sutton (ed.), *Europe and the Caribbean* (London: Macmillan, 1991).

5. M. Cross, *Urbanization and Urban Growth in the Caribbean* (Cambridge: Cambridge University Press, 1979), p. 5.
6. A good account of the political economy of the subregion is provided in J. B. Grugel, *Politics and Development in the Caribbean Basin: Central America and the Caribbean in the New World Order* (London: Macmillan, 1995), pp. 159–95.
7. Membership: Antigua and Barbuda, the Bahamas, Barbados, Belize, Dominica, Grenada, Guyana, Jamaica, Montserrat, St Kitts-Nevis, St Lucia, St Vincent and the Grenadines, Trinidad and Tobago.
8. Membership: Antigua and Barbuda, Dominica, Grenada, Montserrat, St Kitts-Nevis, St Lucia, St Vincent and the Grenadines.
9. Membership: Costa Rica, Guatemala, El Salvador, Honduras, Nicaragua.
10. For a critical review, see C.D. Deere *et al.*, *In the Shadows of the Sun: Caribbean Development Alternatives and U.S. Policy* (Boulder: Westview Press, 1990).
11. For a full account of the evolution of regional integration within the English-speaking Caribbean, see A.J. Payne, 'The Politics of Regional Cooperation in the Caribbean: the Case of CARICOM', in W.A. Axline (ed.), *The Political Economy of Regional Cooperation* (London: Frances Pinter, 1994), pp. 72–104.
12. *The Grand Anse Declaration*, a declaration issued by the Heads of Government of Caribbean Community Countries, mimeo, Grenada, 1989.
13. The West Indian Commission, *Time for Action: the Report of the West Indian Commission* (Black Rock, Barbados: The West Indian Commission, 1992), p. 449.
14. Ibid., p. 446.
15. See *Summary of Conclusions of the First Meeting of the Bureau of the Conference of Heads of Government of the Caribbean Community*, mimeo, 15–16 December 1992, Georgetown, Guyana.
16. For an account of the work of the CBTAG, see the *CBTAG Status Reports*, January and September 1992, Department of State, Commonwealth of Puerto Rico, San Juan.
17. For a fuller discussion, see P.K. Sutton, 'The Banana Regime of the European Community, the Caribbean, and Latin America', *Journal of Interamerican Studies and World Affairs*, 39, 2, Summer 1997, pp. 5–36.
18. This and other subsequent quotes are from *The Convention Establishing the Association of Caribbean States*, Cartagena, Colombia, 24 July 1994.
19. H.S. Gill, *The Association of Caribbean States: Prospects for a 'Quantum Leap'*, The North–South Agenda Papers No. 11, University of Miami, January 1995, p. 9.
20. Ibid., p. 10.
21. *Declaration of Principles and Plan of Action on Tourism, Trade and Transportation*, a statement by the Inaugural Summit of Heads of State and Government and Representatives of the States, Countries and Territories of the Association of Caribbean States (ACS), mimeo, Port of Spain, 17–18 August 1995.
22. Ibid.
23. *Caribbean Insight*, September 1995, p. 1.

7 Going it Alone? The Chilean Strategy for Sub-Regional Integration

Jean Grugel

Chile is the one country in Latin America and the Caribbean (LAC) which is pursuing a strategy of attachment to two subregional groupings. It became an associate member of MERCOSUR, made up of Brazil, Argentina, Paraguay and Uruguay, in June 1997, and is a serious candidate for membership of NAFTA, composed of the US, Canada and Mexico. It has signed free trade deals with almost all other LAC states. Chile is also a member of APEC and is actively seeking to expand trade and investment contacts within Asia-Pacific in a number of other ways. One way to describe Chile's strategy for subregionalism would be 'friends with all but no entangling alliances'. This strategy raises a number of interesting questions about subregionalism in general and within the Americas in particular. Are subregional groupings fixed and ultimately geographically bounded arrangements? Is membership of one subregional group compatible with membership of another? Specifically within the Americas, is MERCOSUR an alternative to NAFTA or are both 'stages' on the road to free trade within the Americas and/or an integrated American bloc? What is implied by membership of both pacts? To what extent are the states engaged in building subregionalist alliances following a consciously planned strategy? These questions form the background to the analysis of Chile presented here. This chapter will argue that Chile's subregionalist strategy is successful precisely because it refuses to be tied exclusively to membership of one subregional bloc in the Americas and combines membership of subregional pacts with a continuing commitment to multilateralism. This suggests that subregional pacts, at least in the Americas, are quite loose and flexible arrangements and that they do not build exclusively on the politics of geographical or cultural identity.

An analysis of the evolution of Chile's subregionalist policies provides an antidote to much of the literature on regionalism in the Americas. Not surprisingly, the main focus of regionalist analysis has been NAFTA.[1] The literature has, on the whole, suggested that subregional integration is primarily an expression of the new basis for US hegemony in the Americas. NAFTA is a reflection of an asymmetrical power relationship and, although the Mexican state is gambling that it will bring investment and trade benefits to Mexico, it undoubtedly involves a heavy cost, which is being paid by the poorest and least privileged members of society and by those geographically removed from the manufacturing and service 'cores' of the border areas and the capital city. The rest of LAC, it follows, can expect to pay similar costs and obtain similar benefits from free trade with the US and Canada as Mexico, only with perhaps fewer net benefits and higher costs. The presumption is that the 'hub-and-spoke' pattern of Mexican integration with the US is the principal model for LAC integration within the Americas. Regional integration is taken to imply a subordination of LAC national policies to those of the US state in exchange for access to the US market.

MERCOSUR has been presented as an alternative model to the NAFTA hub-and-spoke pattern of integration.[2] The difference lies chiefly in the absence of a developed state around which the other members of the pact revolve in MERCOSUR. The politics of MERCOSUR are qualitatively different from those within NAFTA since all members are part of the underdeveloped world, however large some of their economies may be. Chile, however, constitutes a third model for subregional integration in the Americas and, as such, is worthy of analysis. Subregional integration, for Chile, is not a reflection of weakness in the face of global pressures to liberalize, uncertainty about the future prospects of the economy, or a decision to tie Chilean exports into the economy of the US. It has a particular concept of the 'Chilean national interest' at its centre. In Chile, integration is approached by the state from a liberal perspective of the national interest. It is a consequence of the emergence of a state strategy for integration, supported by a coalition of interests from within society.

Chile currently constitutes something of a model within LAC. Not only is the Chilean macroeconomy healthy and its export performance dynamic, but its democracy seems solid and its centre-left government appears to be combining pragmatic policies of economic liberalization with the implementation of social democratic anti-poverty initiatives, funded through socially-pacted tax raises. Regionalism in Chile is thus part of a more general set of development policies from which other LAC states hope to learn. The final part of the chapter, therefore, will assess to what degree the Chilean strategy can be reproduced elsewhere.

THE REGIONAL CONTEXT

Chile was formally invited by the US to open negotiations for NAFTA membership in December 1994. Despite setbacks and disappointments, especially in 1996, these negotiations are still continuing. In June 1996, Chile became an associate member of MERCOSUR, with tariff-free access to MERCOSUR's internal market. Chile's strategy on subregionalism has emerged in the context of a new interest in regional integration across the Americas and the emergence of liberal models of regional integration, in contrast to the protectionist subregionalism of the 1960s and 1970s.

In the period up until the end of the 1980s, the international relations of LAC were determined by the pursuit of nationalist development through import-substituting industrialization (ISI) on the one hand, and the constraints imposed by an international system shaped by the Cold War, on the other.[3] Diplomatically and strategically, LAC states practised a forced loyalty to the US and were staunchly anti-communist in the UN and in the formal organizations of the inter-American system such as the Organization of American States (OAS). The relationship was cemented through a series of bilateral military, defence and aid agreements. This system prevailed, despite a number of important challenges to it from within LAC, until the end of the 1980s overwhelmingly because deviation from it brought retribution and the US's implacable hostility. At the same time, the importance of the bilateral link between LAC states and the US was strengthened because the opportunities for intra-LAC cooperation were limited by the policies of ISI. All states effectively were seeking to industrialize through producing the same range of products (such as industrial goods for domestic consumption, textiles, leather goods, semi-processed food, beer and so on, and, in the larger states, car and automotive assembly and other goods requiring high capital input and/or foreign investment) thereby creating few opportunities for inter-LAC trade and intensifying competition for foreign (generally US) investment.

Not surprisingly, attempts at promoting inter-LAC integration had more of a political than an economic rationale before 1980, namely to challenge US domination and/or to increase the international profile of LAC. A series of subregional integration pacts emerged in the 1960s, such as the Andean Pact and the Central American Common Market. Based partly on an ideological, romantic vision which posited the existence of a Latin cultural world and a common Hispanic legacy, their formal commitment to economic integration was greater than their capacity to deliver it, and they were, inevitably, less than wholeheartedly pursued. The integration process slowed to a halt at the end of the 1970s amid a welter of problems ranging from currency difficulties to political and ideological conflict and straightforward rivalry between members of the pacts. Chile, a member of the Andean Pact on its formation, withdrew in 1974. In any case, by the time of the debt crisis after 1982, this phase of subregionalist integration was effectively coming to a close across the region.

Since the collapse of bipolarity at the end of the 1980s, opportunities for new associations have been created, embodying significantly different understandings of space and boundaries. The result is that alliances of all kinds – security, trade, political – have all been reassessed in the light of new global concerns and systemic change.[4] In particular, the role of the US in the new global order has been hotly debated inside and outside the country. It is clear that from around 1990 US state policies have been intended to create new bases for US hegemony within the hemisphere, although this enterprise is not always supported by popular opinion inside the US or indeed by the US Congress. Partly as a result of new priorities, the US state decided to anchor its economy strongly within the Americas, thereby creating the conditions for a resurgence of regionalism. With the decline of the 'old' security agenda centring on the military, an effort was made to reshape US control of the Americas through trade, development and a commitment to formal democratization. US export-oriented companies, or those that use LAC within their production chains, have sought the support of the US state for their strategy of strengthening their hold on the labour forces and markets of the Americas. NAFTA was signed as a first step, bringing Mexico into a formal integration and free trade pact with the US and Canada. This pushed US relations with LAC firmly into the trade/investment mould and relegated the 'new' security themes of immigration, the environment and drug trafficking to a secondary role. This was a move generally welcomed in LAC, and especially in Chile, as a step towards a liberalized trade environment. The US signalled its interest in creating a free trade area through the Americas and took the lead in setting in motion regular meetings of heads of state as well as organizing bureaucratic and ministerial working groups between American countries with the aim of establishing free trade from Anchorage to Tierra del Fuego by the year 2005. These are known as the Summits of the Americas.[5]

New regionalism in the Americas was liberal in its orientation from the beginning. Protectionism – on the part of the US and Brazil in particular – has sometimes threatened the process but has not succeeded in changing the essential terms under which regionalism now proceeds. Groups in society in favour of protectionism have not been able to gain the upper hand in regionalist negotiations. NAFTA was presented as the first stage in the creation of a free-trade Americas by Washington. President Bush promised, and Clinton reiterated, that NAFTA would be extended southwards provided LAC states put 'their houses in order', meaning successfully implemented economic liberalization. But a number of obstacles currently stand in the way of extending NAFTA to the rest of the Americas, including domestic debates inside the US such as protectionist opposition in Congress and on the part of important Democrat constituencies including organized labour, and Clinton's own political problems. The Mexican crisis of December 1994, in which the US found itself acting as borrower of last resort to the Mexican state, also raised doubts in Washington about the wisdom of taking on formal commitments within LAC; this was the mood of the US Congress which, now with a Republican majority, blocked Clinton's plans of extending NAFTA in 1996. Economic reform has not produced the successes with the required speed in LAC either. And the wisdom of freeing trade too quickly has been questioned, slowing region-wide integration down.

Consequently subregionalist schemes which pull in parts of the Americas have emerged, conceived both as an alternative to NAFTA and as a staging post along the way to the extension of NAFTA southwards. MERCOSUR, which is the main alternative model to NAFTA, is different with respect to the degree and the pace of liberalization and in the relative position assigned to the US within a putatively Americas-wide bloc. But it remains committed to free trade and therefore the existence of two models for integration does not, at least at this stage, present other LAC states with major choices. They are, for the present, compatible models. That they are not competing visions of region-wide integration is crucial for understanding Chile's decision to pursue membership of both subregional pacts.

Subregionalism is further stimulated by the emergence of a common set of political concerns in the 1990s across LAC, as states seek to cooperate in order to strengthen the region's new democracies, all of which have faced internal obstacles and threats of one kind or another. Indeed, democratization in LAC has added a new, and important, dimension to subregionalism in the Americas: an increased protagonism on the part of sectors of civil society in the debate over integration. The business community has been especially prominent, although integration may also create spaces within which other groups from civil society can mobilize.[6] This may eventually undermine or challenge the predominantly liberal and trade-oriented integration schemes currently under consideration and adds another reason to avoid treating the current form of subregionalism in LAC as set in stone.

SUBREGIONALISM AS A STRATEGY FOR ECONOMIC GROWTH

Any explanation of Chile's subregionalist strategy must take the transformation of economic and state structures which were brought about under the dictatorship of General Pinochet (1973–1989) as its starting point.[7] Economic liberalization began early in Chile compared to the rest of LAC. Immediately following the coup, the state was restructured and Chilean capitalism reformed. By the mid-1970s, a uniform tariff of 10 per cent was introduced, state subsidies to industry were withdrawn, labour reform was implemented, a programme of privatization started, covering almost all sectors of the economy except copper, a fixed low exchange rate for the peso introduced and legislation which eliminated restrictions on foreign investment passed. Their swift introduction was facilitated by the disarticulation of the political opposition in the wake of the coup, the repression of labour and of popular organizations, and the enthusiastic support for the reorganization of the economy on the part of business and finance groups. Economic growth was high by the end of the decade but, driven by inflows of external finance rather than exports, the economy crashed in 1981–82 under the weight of a high external debt contracted by private banks, leading to the effective collapse of the domestic financial system when international rates soared, leading to massive levels of unemployment and a wave of domestic bankruptcies.[8]

Second-stage liberalization after 1985 was far less ideological and more pragmatic. The reforms in the 1980s combined liberalization of markets with export promotion and with a judicious role for the state in terms of supporting and encouraging the private sector. Tariffs, which the dictatorship had been forced to raise to 35 per cent in 1984, came down slowly, to 15 per cent by 1988, thereby encouraging export expansion.[9] This was accompanied by a reorganization of the Central Bank, the introduction of tax reforms in 1984 and 1988 (which meant a lighter tax load for business), pension reforms and a second wave of privatizations. Growth, especially in the agricultural sector, took off once again.

As a result, Chile's economy and trade patterns have been dramatically restructured. The economy is now market-driven, export-led and open to foreign investment. Exports have diversified, diminishing dependence on external finance and on copper sales. Export markets have also diversified. By 1992, Chile's most important trading partner was the European Union (EU), with 29 per cent of the country's exports, followed by Japan, with 16.9 per cent. The US, which had been Chile's most important trading partner until the 1970s, had slipped to third place with 16.3 per cent. Sixteen per cent of exports went to the rest of Latin America; Asian countries received 14.2 per cent and 7.7 per cent were exported to a mix of other countries.[10] A diversified export market was one reason why Chile became a model in LAC, where dependence on US markets remains pronounced.[11]

Chile's exports of manufactured and semi-manufactured goods rose by an average of 8 per cent annually between 1980 and 1992, an increase second only to Mexico within LAC.[12] The late 1980s in particular witnessed an increase in the number of firms based in Chile exporting to other countries (national, foreign-owned and joint ventures). The number of firms exporting more than a million dollars worth of goods rose from 235 in 1986 to more than 500 in 1990.[13] As in Mexico, Brazil and Argentina, domestically owned economic groups moved in to fill the vacuum left by the collapse and privatization of state-owned enterprises. Together with new and reorganized branches of transnational corporations, these domestic companies have become 'the agents of the productive model that is being established, particularly in manufacturing and services'.[14]

Of the five major economies of LAC (Chile, Argentina, Mexico, Brazil and Colombia), Chile's dependence on traditional exports fell most dramatically between 1970 and 1994, though it remains heavily dependent on exporting primary goods. Ninety per cent of export firms are located in the primary sector.[15] As a result, Chile remains vulnerable to fluctuations in the international price of copper and other metals and to any changes in the commercial strategies of its trading partners. These could be the result of competition from cheaper producers or from producers of higher quality goods. Chile's new markets for fruit and wine are far from assured. Additionally, the new markets have a ceiling in terms of size which Chile may be close to reaching. The limitations of these new markets account in some measure for Chile's interest in regional integration. A range of industrial, semi-industrial and manufactured goods are exported to Latin America, in contrast to exports to the developed countries. Chilean firms are also more streamlined and competitive than many of their rivals in LAC and could therefore do well as a result of deeper integration. Closer relations within LAC and across the Americas increasingly were seen as means to facilitate export growth.

The resurgence of regionalism at the beginning of the 1990s coincided with the transition to democracy in Chile. The elected governments of Patricio Aylwin (1990–94) and Eduardo Frei (1994–the present) endorsed the policies of pragmatic liberalization and export promotion and set out to develop closer links within the hemisphere. They were heavily influenced by the critiques of the economic policies of the dictatorship which had emerged from within Chile's Christian Democratic think-tanks, especially the *Corporacion de Investigaciones Economicas para Latinoamerica* (CIEPLAN) which concentrated on identifying the means to correct deficiencies in the liberal model rather than on rejecting it wholesale.[16] In other words, CIEPLAN confined itself to technical rather than political critiques. Both Alwyin and Frei represented the *Concertacion Democratica*, the centre-left alliance of the Christian Democratic Party (PDC), the Socialist Party (PS) and the Popular Democratic Party (PPD), a personalist offshoot of the Socialists, in which 'modernizers' were dominant.[17] Without abandoning a commitment to multilateral trade and a unilaterally low tariff, subregionalism was therefore confirmed as an ideal vehicle through which to expand exports within LAC and throughout the Americas as well as to attract investment. Hence subregionalism became a feature of Chile's overall commercial strategy. The vision of subregional integration which emerged from within the new democratic state was, in sum, liberal in conception and orientation and built upon the prior process of economic liberalization under Pinochet.

THE POLITICS OF SUBREGIONALISM: A NEW POLICY COALITION BETWEEN THE STATE AND EXPORTERS

It should be clear by now that, in the Chilean case, the drive to subregionalist integration is underpinned by an ideological rationale that is both economic (opportunities for export growth) and liberal (the prioritization of the private over the public). This approach emerges out of a broader consensus about the overall direction of the economy between actors located inside the state – public officials, the governments of the *Concertacion Democratica* – and some private-sector business organizations. One way to examine this consensus – and therefore explain subregionalism in Chile – is through looking at the relationship between government officials and private-sector business groups over time. This relationship can be conceptualized as a *new policy coalition*. In order to understand its significance, we need to go back to the changes in Chilean capitalism under Pinochet and to the transformation of the left in the 1980s.

Economic change under Pinochet had a profound effect upon civil society. Academic attention has focused heavily upon how liberalism reorganized and disempowered the poor and the working class.[18] However, the capitalist class was also fundamentally reorganized under the dictatorship. Before 1973, Chilean capitalism had been dominated politically by the power of the traditional, landed elite.[19] Import substitution had failed to generate a self-confident industrial elite and copper production, the main export, was in the hands of US multinational companies. Opening the economy prioritized production for export over domestic production. This has weakened Chile's *latifundistas* whose economic interests lie in the production of wine, beef and milk for domestic consumption. It has prompted agricultural diversification into the intensive cultivation of fruit, wood, paper and new mining interests and the modernization of the wine industry. It is also changing business culture by rewarding market-oriented values of hard work, risk-taking and merit over the claims of birth. This has led to the creation of a more dynamic, forward-looking and confident entrepreneurial class.[20] These changes are as yet only partial, but they have created significant cleavages within Chilean business between 'modern' producers and old-fashioned rentiers which are covered over to some degree by the fact that the economic elite remains small and social mobility is difficult. There are significant family and business overlaps between the more dynamic economic groups and the older aristocratic families. The *latifundistas* have not been completely displaced from power. But the changes do threaten the social dominance and political power of Chile's traditional elite. Aristocratic social privilege, upheld by the family networks, kinship and the diffusion of cultural values and symbols which privilege unearned income and family lineage, is – slowly – being eroded.[21]

With these changes, the composition and function of the state altered and state–civil society relationships have been transformed. The dictatorship of Pinochet not only eliminated democratic forms of representation; it also restructured the relationship between the state and socio-economic elites. Economic policy was made by a team of professional economists and right-wing 'technocrats' – better known as the 'Chicago Boys'. The trend was for policy to be to some degree insulated from 'politics', thereby increasing the autonomy of state elites, at least in the economic sphere. However, the economic team under Pinochet was not completely cut off from social pressures. Silva has demonstrated how economic policy was made by an alliance which was formed between the technocrats and key socio-economic groups – businessmen and agricultural producers within the new export industries.[22] This relationship was cemented in place by the very success of the liberal reforms which were introduced; the increased structural weight 'new' capital acquired as a result of the reforms; and the changes in the international political economy more generally pointing to the general adoption of policies of economic openness.[23]

This pattern of state–business relations was an important legacy for the democratic governments.[24] The key departments for making economic policy, of which subregionalism is perceived as part, are Foreign Affairs and the Treasury. Both are staffed overwhelmingly by a new generation of 'technocrats' associated with the *Concertacion Democratica*, drawn in large part from Chile's think-tanks and the universities, now sometimes referred to as 'technopols'.[25] Key positions in the Treasury, the Ministry of the Economy, and in Foreign Affairs, under both Aylwin and Frei, have been held by professional economists who made their careers in international organizations and/or Chile's domestic think-tanks, especially CIEPLAN. These include Alejandro Foxley, Minister of the Treasury under Aylwin; Carlos Ominami, Minister of the Economy under Aylwin; Eduardo Aninat, Minister of the Treasury under Frei; and Jose Miguel Insulza, Foreign Minister under Frei. More junior ministerial posts in the Treasury and the Foreign Office have been filled by Juan Gabriel Valdes (who has been head of the team negotiating with NAFTA and MERCOSUR), Heraldo Munoz, Boris Yopo, Alejandro Jara and Alberto van Klaveren, all of whom held academic positions in the 1980s in Chile and abroad and made important intellectual contributions to the ideological 'modernization' of Chile's centre and left. Democratic state elites now in charge of policy planning for foreign and economic policy recognize the contemporary constraints on government worldwide, and have a sense that the profound restructuring which was carried out in Chile cannot be undone.

As a result, they have pushed for economic and regionalist policies to be made in conjunction with the private sector and they have actively sought to involve export-oriented groups in decision-making. They have sought to work and to consult private-sector groups. At the same time, private-sector business groups have recognized that they need to work with the centre-left government, even if they did not vote it into office, in order to press for the retention of a policy environment favourable to their interests. In other words, a policy coalition has emerged.[26] New policy coalitions are pragmatic and contingent alliances between groups located inside the state and non-governmental elites. They constitute an informal arrangement which brings non-governmental elites into regular and sustained, but unstructured, policy-based discussion with representatives from government, closing these policy areas off from the rest of civil society. In Chile, the formation of a new policy coalition around foreign economic policy-making excludes some social groups (chiefly labour organizations) from influencing policy at the same time as it facilitates the influence of others. As Taylor points out, not all social groups which are organizing in Chile's new democracy have access to the state.[27] Private-business groups are currently privileged in policy-making terms in a number of ways: by the rhetoric of liberal policy-making, by their organizational capacity, by their familial and social contacts with (some) governmental elites and by their growing importance relative to the state in terms of production and export performance.

The origins of the coalition date from the cooperation between the technocrats and export-oriented business groups under Pinochet. However, it is now far more of a strategic and pragmatic relationship than in the earlier period. Export-oriented business and the *Concertacion Democratica* technocrats are not 'natural' allies. Their political loyalties and interests are different.[28] But integration is a priority both for the new, export-oriented businesses looking for markets and for the technocratic elites in the *Concertacion* located in the economic and foreign policy making bureaucracies, seeking growth within the constraints of liberalism. It is seen as a novel way of diversifying exports and deepening Chile's role in the global economy as well as a way to restore Chile's place within the democratic states of the Americas. At least with regard to economic and trade policy – of which regionalism is a part – the state listens to business and business has permeated the policy-making arenas of the state. The result has been the (re)creation of a policy coalition around subregionalism.

In building it, state elites have taken advantage of changes in the patterns of political representation since the transition to democracy. The relationship between capitalist groups and the right-wing political parties, and between capital and the state, has changed fundamentally as a result of the combined effects of economic transformation, the dictatorship and democratization. Before 1973, capitalist elites generally relied on their economic power to pressurize the state, rather than cooperating with it in the formulation of policies. They were loyal to the political party of the right, the National Party, which had not been in power since 1964. After 1973, the links with the political right diminished as parties were banned and capital came closer to cooperation with state elites. In the 1989 elections, for the most part capitalist elites supported Hernan Buchi, ex-Minister of the Economy under Pinochet, partly from ideological conviction and partly because he was regarded as a safe pair of hands on the economy. But since the victory of the *Concertacion*, many within the capitalist elite have accepted the need for cooperation over economic policy although they have little overall ideological sympathy for it. In particular, new business elites sought to take a pragmatic approach towards the government. These entrepreneurs now have the confidence to negotiate directly with the state, bypassing the mediation of the post-dictatorship right-wing parties, the National Renovation Party (RN) and the Democratic Independent Union (UDI), which have been weakened in the process.

The organization of business groups and the way they operate has also changed radically in the course of liberalization. In principle, the business community presents a united front through a number of umbrella associations: the *Sociedad de Formento Fabril* (SFF – the Manufacturers' Association), the *Sociedad Nacional de Mineria* (SONAMI – the National Mine Owners' Society), the *Sociedad Nacional de Agricultura* (SNA – the National Farmers' Society) and the *Confederacion de Produccion y Comercio* (CPC – the Federation of Producers and Traders). There is an attempt to present, at least in public, common positions on policy. In fact, however, there are substantial divisions over policy within Chile's business groups. Business associations cannot agree over a range of economic issues, including trade, protectionism and exchange rates. Pressure for maintaining an open trading environment coexist alongside pressure for protectionism. There are divisions even within the umbrella groups. This means that business relations with the state do not flow only through these groups. Some business sectors have sought to bypass the umbrella groups and develop relationships of their own with representatives of the state. In particular, new businesses, agro-exporters and the manufacturers and suppliers of semi-processed agricultural goods, have been keen to maintain a favourable economic environment for further expansion. These groups have therefore been prepared to work in conjunction with state elites in preparing and negotiating economic policy.

The splits within the economic elite are particularly evident in the key sector for the Chilean economy, agriculture. Formally, the SNA represents all of Chile's farmers.[29] In fact, however, around 80 per cent of its members come from the traditional *latifundista* class, traditionally extremely hostile to the centre and the left. Only 20 per cent of the membership are new exporters and they have only been members since 1990 when the SNA broadened its membership to include smaller farmers. This has meant that the fruit producers have in practice been more active in the *Federacion de Productores de Fruta* (FEDEFRUTA – Fruit Growers Federation) than in the SNA. FEDEFRUTA has been more supportive of the *Concertacion*'s economic strategy – and more influential within it – than the SNA as a whole. The SNA has responded to the preference for protectionism of most of its members and has insisted on defending traditional products, at the expense of the new exporters who can compete in free markets. There have been costs, however, for the SNA in adopting this line. Political decline set in for the SNA with the transition to democracy and shows every sign of continuing. Unlike the new export groups, the SNA has also clung onto its rigid political preferences in social and party terms. It remains closely identified with the right-wing parties. In fact, SNA headquarters functions as a 'social club' for gentleman farmers rather than as an arena for debating agricultural policy. SNA officials have few channels, formal or informal, into the government, including even the Ministry of Agriculture. Its relationship with the Ministry of Foreign Affairs, which has become the key ministry in terms of making policy on regionalism, is distant ('correct but cold', according to one interviewee from the SNA). Unlike representatives of export agriculture, it does not find itself invited to enter discussions with government. Neither President Frei nor his ministers are to be found visiting SNA headquarters and SNA officials are rarely received by the government.

This contrasts sharply with the position of the SFF. Like the SNA, the SFF, which groups together Chilean manufacturing, had played an important role in Chilean politics since its formation.[30] Unlike the SNA, however, the SFF has successfully adapted to the changes in Chile's political economy in the 1970s and 1980s. Despite initial resistance from industries dependent on state subsidies and protected markets, the manufacturing community supported liberalization and benefited from it. It came out of the 1980s strong and confident. Despite ideological differences with the *Concertacion*, the SFF has worked closely with it and formed a 'pragmatic relationship' with the governments of Aylwin and Frei on a range of economic policies, including international trade negotiations, regionalism, some aspects of labour policy and infrastructural modernization.[31] One official from SFF explained that the organization was prepared for problems when the *Concertacion* won the elections in 1989, which have simply not materialized: 'this does not mean that the SFF is ideologically in sympathy with the Concertacion, but that it does not deny it support and advice....and common sets of interest have been built up between the government and the SFF'.[32] The relationship flows through contacts with individuals within the *Concertacion* rather than formally, through institutional channels with particular ministries.

The pragmatic approach of associations such as the SFF and FEDEFRUTA means that policy-making relationships have been built up between state elites and social actors. This is not to say that business can now dictate the terms of economic policy. The authority of the state elites rests on their democratic legitimacy, tied to their technical competence. They feel confident enough to use their specialized knowledge to resist sectoral demands from business where they think appropriate, for example over the unilateral tariff.[33] But the economic team is willing to listen to business, providing that its arguments centre on managing the economic model.

THE POLICY COALITION AT WORK: NEGOTIATING SUBREGIONALISM

Chilean strategies *vis-à-vis* NAFTA and MERCOSUR are determined within the policy coalition. They have gone through important changes since 1990 when subregionalist initiatives were first considered. Policy has shifted from seeing subregionalism as a distraction from multilateralism to the adoption of more complex options which blend a commitment to liberalization with an interest in regional markets. This policy change has been the result of how events external to Chile have unfolded in the course of the 1990s rather than any strategic shift in policy goals. The strategy has, in other words, been adapted as external circumstances have unfolded.

The teams negotiating regionalism have come out of the Ministry of the Treasury (NAFTA) and the Ministry of Foreign Affairs (MERCOSUR). In both cases, the team was led by Juan Gabriel Valdes, reporting first to Eduardo Aninat, Minister of the Treasury over NAFTA and later to Jose Miguel Insulza over MERCOSUR. Alliances between the technocrats and some business groups were important in constructing internal support for both NAFTA and MERCOSUR and in negotiating terms for membership. However, I will show below that the 'technopols' relied on the active support of different business groups in the case of the NAFTA negotiations than for MERCOSUR.

Negotiating NAFTA[34]

Chile was formally invited to open negotiations in December 1994. By early 1995, it was confidently predicted that Chile would become a member the following year. In March 1995, the then US Secretary of Commerce, Ron Brown, even claimed that the negotiations between the

US and Chilean governments were in the final stages, after which the deal would be sent to the US Congress for ratification (*El Mercurio* 28/3/1995). However, in the event, Chile failed to gain accession.

The economic benefits of joining NAFTA were offered as the main reason for seeking accession to the agreement during the 1995 negotiations, although the direct impact that membership would have was in fact slight and often exaggerated in the press. Initially, predictions that exports would increase to the order of 18 per cent were not uncommon. In fact, Chilean elites wanted to join NAFTA for a variety of reasons which ranged from the desire for approval from the US to a wish to liberalize further the Chilean economy, especially public-sector services and government procurement. Attributing liberalization to pressure from the US was seen as a way to depoliticize the debate inside Chile. It was also thought that membership of NAFTA would increase the exposure of Chilean businesses involved in manufacturing to international competition and would contribute to shaping them into a small but highly competitive sector. And finally, accession to the largest trade area in the world would confer on Chile a kind of international 'accreditation', a seal of approval, which would symbolize the success of the economic reforms. As Felipe Larrain of the *Universidad Catolica* in Santiago, claimed: 'it would differentiate Chile from the rest of the region and perceptions of risk that the country represented would fall as a result' (*El Mercurio* 23/3/1995).

Inside Chile, the Treasury was in charge of coordinating the negotiating position. Eduardo Aninat, Minister of the Treasury, held talks with a number of business groups throughout 1995. However, the CPC and the *Camara Chileno–Norteamericano de Comercio* (AMCHAM – the US–Chile Chamber of Commerce) were particularly helpful. Together with the official negotiating team, they coordinated common positions on policy, on how to lobby in Washington (and how it would be paid for), and on what the appropriate national responses might be to the US Trade Representatives Committee (USTR) which was formally in charge of the negotiations from the US side (*El Mercurio* 7/4/1995). The CPC and AMCHAM gave evidence at the International Trade Commission (ITC) in Washington. Additionally, business groups shouldered much of the burden and expense of lobbying. AMCHAM, for example, organized a series of dinners in Washington through the first half of 1995 in order to bring together the presidents of US companies with investments in Chile and the managing directors of US-owned Chilean-based companies, with the US senators and deputies from the US home-state or district where the company was based (*El Mercurio* 7/4/1995).

As the negotiations dragged on through 1995, it became clear that the Republican-controlled US Congress was unwilling to extend the free trade deal negotiated with Mexico through the rest of Latin America as rapidly as Clinton had expected. The negotiations stalled when it became clear that, lobbying in Washington notwithstanding, the Chilean team was unable to convince the US Congress of the value of allowing Chile to join. US officials who had argued for extending NAFTA admitted that many of the problems with the negotiations were unrelated specifically to Chile, and were due to a rising tide of protectionism inside the US (*El Mercurio* 6/2/1996). Consequently by early 1996, the Foreign Minister, Jose Miguel Insulza, admitted that the government was increasingly sceptical about the chances of joining NAFTA in the immediate future (*El Mercurio* 24/1/1996).

Since then, Clinton has continued to repeat his intention of broadening NAFTA to Latin America. Responses in Chile, however, are now more guarded, after the disappointments of 1995–96. There is an awareness that enlargement of NAFTA depends not just on what happens in Chile but also on the internal politics of the US. Insulza is even on record as saying that Chile would not enter discussions until the US administration has Congressional approval for fast-track negotiations (*El Mercurio* 2/4/1996). In the event, however, NAFTA is important enough within Chile that, by early 1997, it was once again preparing negotiating briefs. This time, however, the context of the negotiations is different. In the meantime, Chile has become an associate member of MERCOSUR. This has changed its position within LAC. Also, the expectations about export growth resulting from accession to NAFTA are far more realistic. Consequently, the Chilean negotiators now feel able to make more demands. In particular, Chile is seeking to tighten aspects of NAFTA legislation so that it benefits Chilean businesses. 'Chile should make it a condition that the US makes its rules of origin more flexible in order to make membership of NAFTA more attractive,' chief negotiator Valdes claimed (*La Epoca* 13/12/1996). It may even be that Chile will try to reform both the labour and environmental side agreements to NAFTA, in accordance with the demands of private business, which it sees as a block on its strategy of export-led growth and an attempt by the US to introduce covert protectionism. Chile is also unhappy with aspects of the anti-dumping legislation in its agreements with Canada and would see entry into NAFTA as the opportunity to change this. In all, the negotiating team has become more pro-active in terms of defending national interests.

It is still likely that Chile will be the first new member to be admitted to NAFTA – as soon as the debate about broadening the agreement is resolved inside the US. But meanwhile, the negotiations have been an important learning experience both for the technocrats and for the business associations involved. They reinforced the centrality of multilateralism within the overall commercial strategy of the country, strengthened the cohesion of the international negotiating team and cemented the relationship between the commercial *aperturistas* within the *Concertacion* and the export business groups. As a result, Chile's failure to gain entry into NAFTA in the first round could be said to have had a positive effect on its overall trade strategy. It encouraged Chile to look at other options, including broadening trade deals with Asian countries, membership of APEC and deepening relationships with new trading partners such as New Zealand. A new trade agreement with the EU was also signed in June 1996. In general, it convinced governing and business elites of the importance of developing regionalist policies on more than one front.

Negotiating MERCOSUR

The failure of the NAFTA negotiations led directly to a reappraisal of Chile's relationship with its Southern Cone neighbours grouped together in MERCOSUR. Before any kind of deal could be reached with MERCOSUR, however, a way around the problem of tariffs had to be found since both the technopols and the exporters rejected the use of tariffs as a way of allocating resources and as a device to protect internal producers. In the end, a free trade deal, rather than membership, was agreed in June 1996 and ratified, after some domestic debate, in September of that year.

According to ECLAC economist Hector Asael, the direct economic benefits to Chile of associate membership of this kind are considerable. Not only has access to MERCOSUR's internal markets been achieved, but Chile, by not becoming a full member of the pact, has also been able to protect its economy from unfair competition from Brazil or Argentina should those countries devalue their currencies as a result of economic crisis (*La Epoca* 27/6/1996). It means the introduction of an average tariff of 6 per cent for Chilean goods within MERCOSUR, moving towards completely free trade for most goods by 2003. A list of exceptions, especially in traditional agriculture which will benefit from protection for the first fifteen or eighteen years of the agreement depending on the product, was also negotiated. Tariff-free access to MERCOSUR markets will benefit a range of Chilean goods, including wine and industrial products. The SFF described the agreement as 'of transcendental importance for the country' because it created low-tariff markets for Chile's finished goods (*El Mercurio* 7/9/1996). The Minister of the Economy, Alvaro Garcia, claimed rather grandly that the agreement will lead to 150 000 new jobs in Chile (*El Mercurio* 2/7/1996). Independent analyses were somewhat more sober, suggesting that the increase in exports will be partially offset by a corresponding increase in imports, but pointing out that this will lead to lower consumer prices.[35] And finally, associate membership also leaves Chile free to pursue unilateral trade agreements with other countries.

In contrast to the negotiations over NAFTA, this deal was brokered quickly. The negotiating team on the Chilean side was composed of technocrats from the Ministry of Foreign Relations, led by Valdes once again. In this case, they were supported in particular by the SFF and, in a less obvious way, by FEDEFRUTA. In fact, the SFF saw MERCOSUR as a triumph for Chilean manufacturers and one which it had done much to bring about. The SFF was pleased with the outcome of the negotiations and had supplied many of the technical reports used during the negotiations. In some cases, they were even responsible for drafting parts of the official negotiating documents. The relationship between the government and the SFF during the MERCOSUR negotiations was described in interviews by one official of the SFF as 'one of confidence'. However, ultimately, representatives of both the Ministry of Foreign Relations and the SFF separately concurred with the view that the SFF could only push for particular 'policy options' and shape the agenda: 'in the end the government decided' said one SFF representative. Nonetheless, according to officials within the Ministry of Foreign Relations, no step was taken without consulting export groups. There was a deliberate effort on the part of the government to pull some private-sector groups into the discussions and to win their support.

The deal was rather more controversial inside Chile, however, than NAFTA had promised to be. Despite the benefits for Chilean industry and the support of the SFF, the free trade deal with MERCOSUR opened up a debate because it undermined traditional agricultural products (wheat, beef and milk), staples produced by the large farmers who dominated the SNA. In fact, the divergence of interests within Chile's agricultural community was evident throughout the negotiations. Along with manufacturers, new agro-export groups are eager to keep trade policy biased towards exports. Like the SFF, they were consulted extensively throughout the negotiations. But this did not prevent the SNA trying to block the deal. The SNA actively opposed the deal, claiming that traditional agriculture would suffer losses of around US$460 million as a result of cheaper products coming onto the Chilean market. But the SNA was unable to count on the total support of their 'allies', the two parties of the right in Congress. The RN and UDI split when the vote was taken in Congress and only a handful of representatives finally rejected the agreement (*La Epoca* 14/8/1996). In the end, therefore, the SNA was unsuccessful, although a considerable delay was achieved before free trade would be introduced in wheat and other traditional staples, thereby protecting their markets in the short to medium term.

CONCLUSION: A MODEL FOR THE REST OF LAC?

At the beginning of this chapter, it was pointed out that the success of Chile's economic reforms in conjunction with its defence of a particular set of national interests in approaching subregionalism have turned it into something of a model for other LAC states. Or to put it more precisely, some aspects of Chile's strategy for growth and international insertion are admired by certain LAC elites and by international agencies. The extent to which it is possible to reproduce the Chilean model and whether it is desirable to do so will briefly be considered now.

Subregionalism in Chile is an integral part of a liberal approach to politics, economics and development. Chile's strategy on subregionalism could not be implemented except in the context of liberal politics and economics. It reflects a commitment on the part of state elites to development through free trade and to liberal democracy which sees citizens as individuals and identifies the market as the main mechanism through which goods, services and wealth are distributed in society. This remains fundamentally true, notwithstanding the introduction of social reform and policies to alleviate poverty. Liberal economics have been successful in Chile, broadly speaking, due to a combination of circumstances including their early introduction in the 1970s, geography, the size of the economy, the composition of exports and favourable external support. Few of these conditions can be reproduced by other LAC states.

Additionally, Chile is a stable but limited democracy. Constraints on the extent to which democratization has occurred at the social level have facilitated the prioritization of a liberal pattern of development over redistribution after the end of the dictatorship. It is not surprising therefore that democratization in Chile has been criticized as 'restricted' in that it barely listens to popular movements, non-governmental organizations, women and the poor. The formation of a policy coalition between state elites and export capital around the contours of economic policy has made it particularly difficult for labour or social movements to influence the debate. Once again, it is far from clear that other LAC states could reproduce the combination of circumstances, strategy and luck which have together brought about stable but limited democracy in Chile in which economic liberalism and the prominence of business goes largely unchallenged.

Finally, the implementation of these policies requires an organized and effective state. As Pearce puts it: 'the only general lesson that seems to emerge [from Chile] concerns the importance ... of clearly defined boundaries between the state and the private sector, in which the state promotes the conditions for capitalist development .... This does not necessarily imply a 'weak' or a 'strong' state *vis-à-vis* the rest of society, but it does require a state with *capacity*.'[36] In many LAC countries, the state remains inefficient and its operation patrimonial and clientelistic. This constitutes a barrier to the development of pragmatic and functional relationships with groups in civil society. Also entrepreneurs are frequently weak and lacking in self-confidence and distrust democratizing elites far too much to be able to work with them. So the state may not be able to count on the cooperation of the business community in the same way. Yet if the state is not embedded in society, it is doubtful that it would be able to use subregional integration as a means to enhance development or defend a 'national interest', however it may be defined. In sum, it should be clear that reproducing Chile's political economy is no easy task.

This brings us to the issue of the desirability of adopting the Chilean model. An alliance was established in Chile between the technocrats in charge of economic policy from the *Concertacion* and export-oriented business groups. This strategy is, by its very nature, exclusionary. Not only does it leave out the traditional landowners, who argue for protectionism; it also omits labour and groups representing the poor and the unemployed in policy discussions. Despite the growth of the economy for over a decade now, many Chileans are no better off than they were in the 1970s. Incomes are low, infrastructure is underdeveloped, and the quality of public services, including education, health and social security, is poor. In 1990, per capita income was only US$1750. Income inequalities increased during the dictatorship with around five million of the country's population of thirteen million living below the poverty line by 1989. Little has been done to reverse this since democratization. Even by 1996–97, seven years after the start of the transition, the minimum wage was still only 65.500 pesos (US$156) a month (*La Epoca*, 20 March 1997) and at least 10 per cent of the economically active population were earning less than that. Most people still work in agricultural production, much of which is temporary, and in the unskilled or semi-skilled sectors of the economy. Unionization is low and obstacles to union membership mean that only around 14 per cent of Chile's workforce are in unions. The social costs of this political economy are, therefore, high. Although the new political economy in Chile is based on consensus, democracy and growth do not guarantee economic entitlements for Chile's poor – and nor does the *Concertacion Democratica* suggest that it should. The anti-poverty measures undertaken by the *Concertacion* are of very limited impact. It is not surprising, therefore, that although Chile is an attractive model to political and economic elites because it appears to enjoy both stable government and pro-business policies, many ordinary Latin American people regarded the costs as too high a price to pay.

NOTES

1. For an introduction to the regionalism in LAC, see A. Hurrell, 'Regionalism in the Americas', in A. Lowenthal and G. Treverton (eds), *Latin America in a New World Order* (Boulder: Westview Press, 1994). On NAFTA see S. Saborio (ed.), *The Premise and the Promise: Free Trade in the Americas* (New Brunswick: Transaction Publishers, 1992) and M. Pastor, 'Mexican Trade Liberalization and NAFTA', *Latin American Research Review*, 29, 1994, pp. 153–73. For an updated (liberal) account of NAFTA, see S. Weintraub, *NAFTA at Three* (The Center for Strategic and International Studies: Washington 1997). For a more critical review of NAFTA, see R. Grinspun and M. Cameron (eds), *The Political Economy of North American Free Trade* (New York: St Martin's Press, 1993).
2. The literature on MERCOSUR, especially in English, is scarce. See R. Barbosa, 'O Mercosul e suas institucoes', *Boletin de Integracao Latino-Americana*, 14, 1994, 1–2 and R. Lipsey and P. Meller (eds), *NAFTA y MERCOSUR* (Santiago: CIEPLAN, 1996). On Brazil in MERCOSUR, see M. Gomes Saraiva, 'La politica exterior brasilena. En busqueda de un paradigma', *MERIDIANO-CERI* (Madrid, June 1997, 15, pp. 22–7).
3. J. Nef, 'The Political Economy of Inter-American Relations', in R. Stubbs and G. Underhill (eds), *Political Economy and the Changing Global Order* (London: Macmillan, 1994).
4. See J. de Melo and A. Panagariya (eds), *New Dimensions in Regional Integration* (Cambridge: Cambridge University, Press, 1993) and, L. Fawcett and A. Hurrell (eds), *Regionalism in World Politics: Regional Organization and International Order* (Oxford: Oxford University Press, 1995).
5. See A. Payne, 'The United States and its Enterprise for the Americas' and J. Grugel, 'Latin America and the Remaking of the Americas', in A. Gamble and A. Payne (eds), *Regionalism and World Order* (Basingstoke: Macmillan, 1996), for a more detailed explanation.
6. Cook, 'Regional Strategies and Transnational Politics: Popular Sector Strategies in the NAFTA Era', in Chalmers, Vilas *et al.*, *The New Politics of*

Inequality in Latin America (Oxford: Oxford University Press, 1997). See also J. Grugel (ed.), *Democracy without Borders: Transnational Actors in Democratic Consolidation* (London: Routledge, 1998).
7. This section draws upon J. Grugel, 'The Chilean State and New Regionalism: Strategic Alliances and Pragmatic Integration', in J. Grugel and W. Hout (eds), *The New Regionalism: Across the North–South Divide* (London: Routledge, 1998 forthcoming).
8. L. Oppenheim, *Politics in Chile: Democracy, Authoritarianism and the Search for Development* (Boulder: Westview Press, 1993).
9. R. Saez, 'Estrategia comercial chilena. Que hacer en los noventa?', *Estudios CIEPLAN*, 40, March 1995, pp. 21–38 and R. French-Davis and R. Saez, 'Comercio y desarrollo industrial en Chile', *Estudios CIEPLAN*, 41, December 1995, pp. 67–98.
10. Saez, op. cit.
11. ECLAC, *Open Regionalism in Latin America and the Caribbean. Economic Integration as a Contribution to Changing Production Patterns with Social Equity* (Santiago: ECLAC, 1994).
12. ECLAC, *Policies to Improve Linkages with the Global Economy* (Santiago: ECLAC, 1995), p. 74.
13. P. Meller, *Un Siglo de Economia Politica Chilena (1890–1990)* (Santiago: Editorial Andres Bello, 1996) p. 275.
14. P. Gerchunoff, *Las Privatizaciones en la Argentina Primera Etapa* (Buenos Aires: ECLAC/Instituto Torcuato di Tella, 1992).
15. R. Laban and P. Meller, 'Estrategias alternativas de comercio para un pais pequeno: el caso chileno', in R. Lipsey and P. Meller (eds), *NAFTA y MERCOSUR*, op. cit.
16. G. Arriagada and C. Graham, 'Chile: Sustaining Adjustment during the Democratic Transition', in S. Haggard and S. Webb (eds), *Voting for Reform: Democracy, Political Liberalization and Economic Adjustment* (Oxford: Oxford University Press, 1995).
17. On the modernization of Chilean politics, see R. Barros, 'The Left and Democracy: Recent Debates in Latin America', *Telos*, 68, 2, 1986 pp. 49–70 and M. Garreton, 'Political Democratisations in Latin America and the Crisis of Paradigms', in J. Manor (ed.), *Rethinking Third World Politics* (London: Longman, 1991).
18. See Oppenheim, op. cit.
19. B. Loveman, *Chile* (Oxford: Oxford University Press, 1979).
20. C. Montero, *La Revolucion Empresarial Chilena* (Santiago: CIEPLAN, 1997). See also E. Bartell, 'Perceptions by Business Leaders and the Transition to Democracy in Chile', in E. Bartell and L. Payne (eds), *Business and Democracy in Latin America* (Pittsburgh: University of Pittsburgh Press, 1995).
21. Ibid.
22. E. Silva, 'Capitalist Coalitions, the State and Neoliberal Restructuring: Chile 1973–1988', *World Politics*, 45, 1993, pp. 526–59.
23. M. Grindle, *Challenging the State. Crisis and Innovation in Latin America and Africa* (Cambridge: Cambridge University Press, 1996).
24. This section relies on interviews carried out by the author with representatives of the Chilean Foreign Office in March–April 1997.

25. E. Hershberg, 'Market-Oriented Development Strategies and State–Society Relations in New Democracies: Lessons from Contemporary Chile and Spain', in Chalmers, Vilas *et al.*, op. cit.
26. For a more detailed discussion of new policy coalitions, see J. Grugel, 'State and Business in Neo-Liberal Democracies in Latin America', *Global Society*, 1998 (forthcoming) and J. Grugel and A. Payne, 'Regionalism, Development and the State in the Caribbean Basin', in B. Hettne and A. Iotai (eds), *National Perspectives on the New Regionalism in the South* (Basingstoke: Macmillan, 1999).
27. L. Taylor, *Citizenship, Participation and Democracy: Changing Dynamics in Chile and Argentina* (Basingstoke: Macmillan, 1998).
28. This section draws on interviews with representatives of the *Sociedad de Fomento Fabril* (SFF) in Chile in March–April 1997.
29. The following paragraphs draw on material from interviews with representatives of the *Sociedad Nacional de Agricultura* (SNA), Santiago, March–April 1997.
30. M. Cavarozzi, *The Government and the Industrial Bourgeoisie in Chile*, Unpublished Ph.D. thesis, University of California, Berkeley, 1973.
31. This was the term used by a representative of the SFF in interviews.
32. Ibid.
33. Interview in PROCHILE, the commercial promotions department of the Foreign Office, March–April 1997.
34. This and the following sub-section draws extensively on interviews conducted with Foreign Office, SFF and SNA representatives, as well as accounts from the Chilean press. Press sources are cited in the text.
35. E. Muchnik, L. F. Errazuriz and J. I. Dominguez, 'Efectos de las asociacion de Chile al MERCOSUR en el sector agricola y agroindustrial', *Estudios Publicos*, 63, 1996, pp. 113–64.
36. J. Pearce, 'Chile: Democracy and Development in a Divided Society', in A. Leftwich (ed.), *Democracy and Development* (London: Polity Press, 1996).

Part 3
Subregionalism in East Asia

Introduction

by Glenn Hook

The initiatives to promote subregionalism in East Asia have their origins in the history of the region as well as in the more immediate global transformation engendered by the ending of the Cold War, the strengthening of regionalism elsewhere, as in Europe, and the promotion of a larger regional project seeking to bind together the Asian and Pacific wings of the global political economy. The different form that subregionalism takes in East Asia thus stems from historical structures as well as the contemporary changes generating intra- and extra-regional pressures on the East Asian states and economies in the context of globalization and regionalization processes. In the face of moves seen as possibly leading to 'closed regionalism' in other parts of the world, the state and economic elites of the region have sought to ensure the survival of their mainly export-oriented economies by promoting new subregional initiatives, as with the East Asian Economic Caucus (EAEC), at the same time as older subregional organizations have been reinvigorated, as in the case of the Association of Southeast Asian Nations (ASEAN). In this sense, the development and strengthening of the regionalist project in Europe, along with the launch of the region-wide Asia Pacific Economic Cooperation (APEC), can be said to have stimulated subregional responses in East Asia.

Historically, the major motivation for the creation of the key subregional organization, ASEAN, was considerations of security more than economics, with the need to put an end to intraregional conflict seen as the basis for economic development and nation-building. Before subregionalism could sink any roots, therefore, not only the sources of conflict, but conflict itself needed to be addressed. The end of the 'Confrontation' between Indonesia and Malaysia in the early 1960s is testimony to the ability of the newly formed states of Southeast Asia to help to bring about stability in this subregion. Its role in resolving the war in Cambodia, along with the Association's recent promotion of the ASEAN Regional Forum (ARF), point to its continuing security concerns. Now, this is not so much in terms of seeking to dampen intraregional conflicts, though these still exist, but in ensuring the engagement of the big powers in the new multilateralism at the heart of the ARF process. In what way the big powers continue their involvement in the region will remain central to the development of subregional initiatives in East Asia.

Indeed, subregionalism in East Asia has been complicated by the involvement of the big powers. The role of the big outsider in the hot wars in Korea and Vietnam is not the only concern of the leaders struggling to overcome the legacy of Western imperialism in pursuing their goal of economic development and prosperity. The classic regional order of China and the attempt to impose a regional order through the Empire of Japan's 'East Asia Coprosperity Sphere' suggest why the weak subregional states are sensitive to the intraregional big powers, too. Whether as part of a Chinese, Western or Japanese-imposed order, the peoples of Southeast Asia have been in a subordinate position, with few means to shape their own destiny. This helps to explain some of their resistance to the US proposal in the 1950s to establish the Southeast Asian Treaty Organization (SEATO), their long-standing concern about the involvement of Japan in any defence arrangements, and their perennial sense of vulnerability in the face of China. With China as a neighbour, an intraregional 'balance-of-power' approach to security has not been a viable option for these states, especially when the 'Chinese threat' could as much come from ethnic Chinese within state borders as the giant beyond.

The economic development of China as well as Southeast Asia in recent years has made economic subregionalism the most pronounced aspect of subregional developments in East Asia. In the latter case, this can be seen in the Southeast Asian efforts to push forward with the ASEAN Free Trade Area (AFTA) and to maintain a voice in the APEC process. In the case of China, the networks of Chinese in the region are much more interested in *renminbi* (currency) than revolution. The ethnic ties uniting the Chinese have been the basis for making investments, forging economic links, and creating a new division of labour in the region. Tie this to the politics of identity, and the idea of a 'Greater China' subregion composed of Hong Kong, Taiwan and Southern China starts to take on economic and political meaning in the emerging political economy of East Asia. It is the attempt to invent such a subregional identity which helps to explain some of the tensions and contradictions at the heart of subregionalism in East Asia.

A subregion based on Chinese networks is a different form of subregionalism from the proposal made by Prime Minister Mahathir of Malaysia, whose attempt to establish the EAEC in 1991 followed the moves to establish the European Union (EU) and the launch of the APEC. It is a subregional initiative which seeks to pull the economic big power of the region, Japan, into playing a more pro-active role as the voice of Asia in the arenas of the global political economy. In a sense, the Japanese transformation of the region through investments, transfer of manufacturing production, and the creation of a variety of production networks made such a subregional economic initiative seem viable as a

Introduction: Subregionalism in East Asia 167

way to develop an 'East Asian' as opposed to an 'Asia Pacific' or 'Pacific' identity. The unwillingness of Japan to offer support for the Malaysian proposal in the face of US opposition points to the continuing power of the bilateralism at the heart of the Cold War regional order to constrain the behaviour of states in the region, with Japan being locked into the Pacific at the same time as the pull towards Asia is emerging. By following the US in support of APEC as a wider regional project seeking to push liberalization and market-led answers to the question of development, the EAEC remains as a political as well as an ideological challenge to the big power's attempt to restructure the regional political economy.

The newly invigorated ASEAN, 'Greater China' and EAEC are three of the main subregional projects being pursued in post-Cold War East Asia. Their history, evolution and role help to explain some of the general trends towards subregionalism in East Asia, which have emerged in response to both regional and global trends. The three chapters making up this section deal with each of them in their historical and contemporary setting. The discussion should make clear that, for all their diversity, these subregional initiatives have taken place within the context of the structural constraints imposed by historical legacies, on the one hand, and the processes of globalization and regionalization, on the other.

8 The Association of Southeast Asian Nations

Dominic Kelly

The Association of Southeast Asian Nations (ASEAN) can be regarded as the key subregional grouping in Southeast Asia. It was founded in the city of Bangkok on 8 August 1967 by agreement between Indonesia, Malaysia, the Philippines, Singapore and Thailand. They were joined on 8 January 1984 by Brunei and on 28 July 1995 by Vietnam, with the addition of Laos and Myanmar on 23 July 1997 almost completing the long-cherished aim of an 'ASEAN-10'. Only the absence of Cambodia stands in the way of this goal but its membership, scheduled for the same day as Laos and Myanmar, was delayed following the coup led by the Second Prime Minister Hun Sen. The goal of the organization, enshrined in the Bangkok Declaration of 1967, was summarized and made explicit in 1983 by the then Prime Minister of Thailand, Prem Tinsulanonda, in his opening address to the 16th ASEAN Ministerial Meeting when he stated categorically that ASEAN 'stands for peace and prosperity for Southeast Asia'.[1] Two points of immediate interest may be drawn from this statement. First, it is clear that prosperity is understood as a mechanism through which to achieve peace and stability, both within individual ASEAN states and amongst members of the organization. Economic growth is apparently primary and central in this regard. Second, is a belief that what can be done within ASEAN can be achieved elsewhere in Southeast Asia. Intra-ASEAN economic cooperation is thus not concerned simply with wealth creation but has wider social and political goals. In combination, these two points lead us to surmise that ASEAN is to be understood not just as a structure, the institutional agglomeration of member state interests, but as a *process* redolent with meanings and aspirations. In this chapter we explore this notion of ASEAN as a process rather than as a static ensemble of states and state interests. Far too often, an understandable interest in the institution of ASEAN obscures rather than reveals the power relationships inherent within it. In this regard, the signal failure of Prem Tinsulanond to make any mention of, let alone connection between, prosperity and democracy is particularly marked. ASEAN has not only failed large sections of its collective population in economic terms but, so too, has it failed to promote the cause of democratic accountability. It may well, then, stand for peace and prosperity for Southeast Asia, but only for some.

The economic agenda of ASEAN has, however, always been subordinate to the fulfilment of a more pressing security agenda, to the extent that the former has until recently been almost vestigial in aspect. According to Leifer, ASEAN was created primarily 'to provide an institutional framework for intra-regional reconciliation and to establish a corresponding trust among former adversaries following Indonesia's practice of Confrontation'.[2] Part of our task is to explore these origins. Nevertheless, what was true then does not necessarily hold now, and another part of our task is to outline and account for ASEAN's reaction to the end of the Cold War, the relative decline of the United States, the dominance in the subregion of Japanese economic interests and the growing presence of the People's Republic of China (PRC). At a certain analytical focus, the ending of the Cold War, increasing trade friction between the US and Japan and the liberalization of the Chinese economy are indicative of the increasing salience of economic issues on the agenda of regional politics and diplomacy. At a wider focus they are symptoms of fundamental changes in the global political economy, the consequences of which have not yet been fully understood. Whatever the focus, however, it is clear that separately and collectively the ASEAN states are in a position of relative but still fundamental weakness in the global political economy, and will remain so for some time to come. Accordingly, in the face of continual external interference and pressure by colonial, 'great' and 'super' powers the social, economic and political history of ASEAN has and continues to be characterized by the need to fight for the right to manage its internal and external affairs. It is this struggle which forms the backdrop to the discussion that follows.

The first section of the chapter sketches an overview of what appear to be the important socio-economic, politico-military, ideological and intellectual movements and contradictions running through and around the ASEAN subregion. The second section covers the origins and early history of ASEAN within the context of post-War US hegemony, while the third and fourth seek to bring this into the present by outlining contemporary economic and military-strategic developments in light of the end of this hegemony.

OVERVIEW

As noted above, the immediate origins of ASEAN are to be found in the military-strategic sphere, and can be traced to events at the global, regional and national/local levels during the 1960s and before, although Leifer's comments notwithstanding, the lines and direction of causation operating within and between these are so complex that it is well-nigh impossible to unravel satisfactorily the decisive element. Nevertheless, it is clear that the major elements were the global confrontation between the United States and the Soviet Union which had arrived in Northeast Asia in the years leading up to the Korean War in 1950 and in Southeast Asia following the withdrawal of French forces from Vietnam in 1954; the perception of the increased ideological and military threat emanating from a radicalized China in the throes of the Cultural Revolution and (from 1967) in possession of nuclear weapons; the ending of the regional confrontation between Malaysia and Indonesia (1963–66); and the various irredentist, ideological, ethnic, racial, historical and geographical problems and anxieties commonly felt but uniquely experienced during the struggle to establish, maintain and embed national independence from the former colonial powers.[3]

A shared sense of vulnerability to external and internal threats, therefore, provided the glue which bound the original members of ASEAN together. This glue held throughout the 1970s and 1980s, its strength maintained by ASEAN-wide fears pertaining to the longevity and resurgence of the Cold War conflict, the local pretensions to dominance of Vietnam and persistent concerns over domestic insurgency and unrest.[4] The members of ASEAN remain open to this sense of vulnerability in the post-Cold War 1990s, despite the waning of ideological confrontation and years of solid economic growth.[5] The absence of Soviet/Russian forces, the retrenchment of the US and the growing confidence and assertiveness of the People's Republic of China has in some respects brought more rather than less insecurity as the traditional Great Powers of the region, China and Japan, loom ever larger on the horizon. To these concerns must be added the absorption into the ASEAN process of Vietnam, Laos, Myanmar and Cambodia, all with distinct economic systems and political traditions; realignments among security, economic and racial/cultural issues and consequently political agendas, particularly as regards the emergence of an Asian middle class more and more demanding of the democratization of systems of government ranging from plutocratic authoritarian to military dictatorships; and understanding and coping with the intensification of economic competition associated with the globalization of economic activity.[6] Finally, despite its undoubted success as a tool of confidence-building, the ASEAN-wide predilection for 'quiet' diplomacy has served not to solve some of the long-standing intra-ASEAN problems but to place them on the proverbial back-burner where they simmer to this day.

Meanwhile, with the addition of three new member states and the prospect of a fourth, ASEAN has at long last proved itself confident enough to embrace a process of expansion indicative of its transition from a subregional to a regional organization. ASEAN thus seems suddenly more disposed to become involved in the big political and economic issues, a major departure from its previous interest in avoiding high politics in favour of state-building and regime maintenance through the mechanism of economic growth. Indeed, while ASEAN has performed an important peace-maintenance and confidence-building role with regard to its members, with the exception of the Cambodian peace settlement this has been almost exclusively intra-ASEAN and limited to specific problems. In contrast, the creation of the ASEAN Free Trade Area (AFTA) and the ASEAN Regional Forum (ARF) are major developments in the economic and security fields. Moreover, the unfolding of an ideological project on the scale of the East Asian Economic Caucus (EAEC) is politics of an altogether different stripe, challenging as it does the dominant conflictual interpretation of the pattern of international relations and threatening to change the future shape and direction of the regional order.[7] In combination, these developments signal an attempt by ASEAN to create and protect the social, economic and political space it needs to survive into the new millennium. In the meantime, however, issues of poverty, injustice and repression, extant in one form or another throughout the ASEAN states, have been greeted by an almost deafening silence, raising questions as to the extent and depth of the so-called East Asian 'miracle' and the interests of those who perpetuate the myth of its existence.[8]

Central themes

With regard to these changing circumstances, two closely related themes seem to be of particular interest. The first revolves around the search for understanding the success of ASEAN in light of the acknowledged failure of so many other attempts at institution-building in East Asia in the post-World War Two era. Past attempts include the US-sponsored Southeast Asia Treaty Organization (SEATO), established in Bangkok in 1955 and including amongst others Thailand and the Philippines as members; the Anglo-Malayan Defence Agreement of 1957 which by 1971 had metamorphosed into the Five-Power Defence Arrangements with Britain, Australia, New Zealand, Malaysia and Singapore as members; the pro-western Association of Southeast Asia (ASA), established in 1961 between Thailand, Malaya and the Philippines; and the Malay-oriented Maphilindo, established in 1963 between Malaysia, the Philippines and

Indonesia. Although we must not over-generalize since each was set up for different though often cross-cutting reasons and goals (hence the frequent overlap in membership), what sets them apart from ASEAN is that all have failed or, at the very least, failed to live up to expectations.[9] Against this background of relative failure, ASEAN's achievement has been attributed most often to apparently unique features of the organization associated with what we have for the sake of convenience thus far called 'quiet' diplomacy: namely its strict adherence to informality and the avoidance of excessive institutionalization, the pursuit of consensus through non-hostile negotiation, and the pursuit and tolerance of bilateral diplomacy between members in the context of multilateralism.[10] Arguably, however, these or similar features have variously appeared at other times in other settings, which in combination with the historical circumstances surrounding the creation of the organization points us to the conclusion that the aim of quiet diplomacy is simply to raise the 'comfort level' between participants. In accepting that this is indeed the case we have good reason to ask why so much ink has been spilled trying to explain quiet diplomacy: what other possible purpose(s) does it serve? Two answers spring immediately to mind. First, that promotion of supposedly unique and therefore irreplaceable attributes serves to differentiate ASEAN from the West; and second that it deliberately and self-consciously excludes the general populace of ASEAN from the decision-making process. In effect, ASEAN as both institution and process is exclusionary, unaccountable and opaque. More charitably, a third answer, and one that does not negate the other two, may be that promotion of the uniqueness of the cultural attributes and values supposedly inherent in quiet diplomacy serves to preserve and extend the shelf-life of a relatively powerless subregional organization caught between the hammer and anvil of Great Power interests.

The second theme revolves around the dynamic tension created by the promotion of quiet diplomacy as the concrete manifestation of an amorphous but shared Asian historical and cultural/racial heritage, an Asian or ASEAN or Pacific 'way', and the encrustation of the (purportedly) consensual, informal ASEAN process with the more confrontational, formal, legalistic and generally weighty institutional structures and practices associated with the functionalism of the European Union (EU), in particular, and western multilateral institutions in general.[11] This encrustation, embodied and manifest by the advent of the AFTA and the ARF in 1992 and 1994 respectively, has been accompanied and supplemented by a sudden proliferation of larger regional organizations which both embrace and overlap Southeast Asia, including the inauguration of the Asia Pacific Economic Cooperation (APEC) in 1989 and the Asia–Europe Meetings (ASEM) in 1996.[12]

Arguably, then, the inclusion of new issues and participants will lead to more formality, institutionalization and confrontation in negotiation amongst members, resulting in the erosion of the foundations of quiet diplomacy upon the roots of which the success of ASEAN apparently rests. This is all well taken. However, if quiet diplomacy serves an ideological as well as a practical purpose (and there is no doubt that it has served a practical purpose in reducing tensions and increasing confidence between member states) then the whole debate over a distinct ASEAN way really masks something more fundamental. To my mind, this is that propagation of the ASEAN way is part of a political project that itself is a direct response to the end of US hegemony and its subsequent well-documented recourse to unilateralism. Thirty years of relative disillusionment with its allies has seen the US grow increasingly attracted to the pursuit of unilateralism in both the economic and military-strategic spheres, and this has impacted on East Asia as elsewhere. The US has shown itself less willing to tolerate policy difference in the region, and then only so long as this did not conflict with its larger purpose (that is, a peaceful capitalist ASEAN as a bulwark against communist Indochina). With the end of the Cold War, US toleration of difference – economic, political, strategic and/or social – may evaporate, and perhaps already has. The negative US response to the promotion of distinct definitions of human rights as a central component of the ASEAN way may be cited as direct evidence of this, which the continued reluctance of successive administrations in Washington to punish Chinese human rights violations only serves to confirm.[13] Great Powers, it would seem, continue to play under different rules. That the US has been prepared to modify its traditional preference for bilateral relationships by participating in the ARF and APEC can in this light be viewed as part of wider, post-hegemonic efforts at imposing new rules of the game on ASEAN and others. In the absence of a recourse to economic pressure and/or other forms of resistance, promotion of the ASEAN way has been the response, and one that pre-dates the end of the Cold War.

Accepting uncritically the elaboration of the two themes as outlined above, the implication is that they are in the process of colliding with one another. To recapitulate, the success of ASEAN, achieved by dint of the unique cultural attributes with which it is imbued untainted by insidious western influences and values, is threatened by the encroachment of those self-same western influences and values from within and without by virtue of the fact that ASEAN has consciously accepted responsibility for tackling issues and problems involving non-Asian countries which the organization was never meant to solve, and by its acceptance of the discipline of the global economy which cannot fail to do other than erode and destroy established economic and social patterns and relationships. Accordingly, the survival of the organization in its present form is called into question. If, however, the success of ASEAN is in fact not entirely due to or underpinned by cultural attributes and qualities unique to the organization but to the historical circumstances and opportunities extant at the time of its creation, circumstances ultimately shaped by failing US hegemony, then this line of argument does not necessarily hold. It falls to us, therefore, to re-examine with greater care why it is that there is such a strong ideological component to the ASEAN process, and whether, how and why this might be changing in the post-Cold War era. Does the formulation of an ARF based, at least rhetorically, on Asian rather than western principles and values say something about the changing nature of global politics manifest in the rise of an alternative vision or form of socio-political and/or economic organization, or is it simply a last-gasp effort by the ASEAN states to remain close to the diplomatic centre? Both suggestions contain more than a grain of truth, and obviously feed and bounce off one another. ASEAN is as much a child and victim of historical antagonism, territorial disputes and clashing ethnic and religious interests, as well as irredentism and economic dependence and vulnerability as it is of Cold War bipolarity. So too, therefore, is the ASEAN way.

In turning to an historical analysis, then, what we require is not so much an understanding of the *institutional* growth of ASEAN, but of the emergence and maturation into political discourse of the ASEAN way; one that attempts to collapse the distinction between 'local' and 'global' explanations.[14] In order to achieve this, however, we must first gain an understanding of events at the global level in the aftermath of the Second World War, and of the peculiar role that Japan was to play in them.

ORIGINS OF THE ASEAN WAY

In the eyes of US planners the arrival of the Cold War in Asia prompted a fundamental reappraisal of existing policy towards the region. For Southeast Asia the end result was to be its effective subordination to US hegemony demanding, firstly, renewed access for Japan to the raw materials and markets of its former Southeast Asian colonies and, secondly, the insertion or reinsertion of Southeast Asia into the international economy.[15] The implications of this reading of history are immediately clear: ASEAN has from its inception been locked into a capitalist path to development dominated by the economic requirements of the former colonial power, Japan, itself subordinate to the hegemonic interests of the US. We may suggest from the outset, therefore, that the position ASEAN occupies within the global political economy is one of fundamental weakness and dependence, even though this may be relatively less than that suffered elsewhere. Neither the small latitude in choice of political settlement afforded to ASEAN elites under US hegemony nor the spectacular economic growth rates enjoyed during the late 1980s and early 1990s can completely hide this fact.

Japan's new economic role was to be pro-active. The overriding aim was to create a 'second rank' economy strong enough to withstand Communist subversion from within and, with US military backing, the Soviet threat from without. A secondary but related aim was for Japan to act as an engine of growth for the non-communist countries of Southeast Asia through procurement of raw materials and the infrastructural investment needed to facilitate this. Symbolically, post-militarist Japan was to act as a model and exemplar to other states in the region of the intertwined benefits of a capitalist economy and of a liberal democratic polity. While these plans were being laid, the Korean War provided a major boost to the rebuilding of Japan's economic strength and its first post-war contact with an Asian country.

Meanwhile, many of the reforms considered essential to the democratization of Japan were either reversed or simply not implemented.[16] Key here is the US-sponsored rehabilitation of the major Japanese corporations within an economic, social and political framework broadly supportive of their needs, and their unique adaptation of the Fordist mode of production in keeping with local circumstances.[17] This adaptation, undertaken so as to satisfy the needs of the Japanese elite at the expense of the mass of the population, provides perhaps the supreme irony of what was initially an almost laudable attempt by the US to expunge militarism from Japanese society and inculcate in its place democratic institutions, practices and ideals. So desperate were the Occupation authorities and their superiors in Washington to succeed in this monumental undertaking that they were prepared to hold the experiment up as a success even after domestic pressure from US business interests in Japan, the resistance of the Japanese elite and the onset of the Cold War had corrupted it almost totally. Thus, far from being a model and exemplar to other states in the region of the intertwined benefits of an idealized capitalist economy and liberal democratic polity, Japan has become a model of something almost completely different: an authoritarian society capable of delivering high and sustained economic growth for the benefit of a small section of the population at the expense of the majority. It is this model that certain members of ASEAN, chiefly but not exclusively Malaysia and Singapore, have so enthusiastically embraced and upon which the ASEAN way has been predominantly based.

If the US failed in its attempt to fully democratize Japan it succeeded beyond all hopes in its aim of securing Japanese economic access to non-communist Southeast Asia.[18] The attributes of Japanese society and economy outlined above, in combination with what for Japanese business and government was a global and regional environment ripe for exploitation, provided the means and impetus for the five decades of Japanese penetration of the region that followed. In addition to the massive problems of environmental pollution, resource depletion and exploitation of the working population, the consequence of this penetration has been, arguably, the domination of the productive process in the ASEAN subregion (and, potentially, of Southeast Asia as a whole) by material and ideational forces closely linked to if not synonymous with Japan.[19] Speculation as to the eventual consequences of this process is just that, speculation. Nevertheless, if, as some argue, a Japanese-centred regional bloc does emerge it will be rooted in production, in the world of work, not in patterns of exchange (of goods or services) no matter how large or dense.[20]

More important to our present purpose, however, is to acknowledge the debt that the proponents of the ASEAN way owe to Japan's 'miracle' economy and the various lessons drawn from it to create a model simple enough not only to be copied but transformed into an ideological tool with which to build notions of Asian exceptionalism. These notions not only seek to differentiate East from West but are supportive of a model of development which works 'to naturalize the authoritarian institutional and social relations of the East Asian Miracle'.[21] They thus contain within them the potential to unite and perhaps sustain ASEAN- and Asian-wide resistance to US and European demands for economic and social conformity in the context of post-Cold War capitalist triumphalism. On the other hand, notions of Asian exceptionalism also contain the seeds of contradiction and resistance, at both the popular and elite levels, by virtue firstly of their reliance on imagery associated inevitably with the Greater East Asia Co-Prosperity Sphere and, secondly, of the social contradictions inherent in a model which seeks explicitly to subjugate the needs of the individual to the priorities of community and society. To take just one example from Singapore, in order to further the cause and institution of 'patriarchy' in that country, legislation was passed in 1994 making adult offspring legally responsible for the support of their parents, refusing medical benefits to the families of female civil servants and disallowing subsidized housing to unmarried mothers.[22] The potential for resistance to these measures could not be clearer. Moreover, as Berger's analysis clearly shows, the many voices raised in praise of Asian exceptionalism sound a note not of harmony but of discord, with some (for example, Malaysia) drawing on a Japanese and others (for example, Singapore) a Chinese model simultaneously linked and riven by a nebulous Confucian philosophical heritage.

Whether or not a Japanese-centred regional bloc does emerge, and whether or not such a bloc is rooted in economic or ideological linkages or a mixture of both, the fact remains that this relationship of Japanese dominance and Southeast Asian dependence was forged and tempered within the wider context of US global hegemony. By the late 1960s when ASEAN was created this hegemony was in question; by the early 1970s it was over. Evidence of the incipient loss of hegemony was there for all to see in 1969 when the Nixon 'Doctrine' shifted the onus for the defence of the non-communist world away from the US and towards its allies, and was confirmed by the devaluation of the dollar two years later. The underlying message was clear: from this point on the US would no longer put the stability of the world order before its own immediate interests.

For ASEAN, these developments presaged major changes. The announcement of the Nixon Doctrine and the eventual US withdrawal from Vietnam brought with it a reorientation of ASEAN security concerns outwards, most importantly to fill an immediate need to secure itself from the danger of conventional attack by a reunified Vietnam, whose regional ambitions were fuelled and set alight by the transformation of the conflict in Southeast Asia from one of East–West to one of East–East as China and the Soviet Union fought a proxy war in Cambodia for supremacy in the Communist world. In response, ASEAN attempted to protect its external borders in several ways, including absolute support (in public at least) for Thailand in its role of front-line state; quiet encouragement of a greater Japanese diplomatic presence in the region and effective use of the United Nations in securing condemnation of the Vietnamese-installed government in Cambodia.

Coping with the Cambodian crisis added the institutional focus crucial to the propagation of the ASEAN way and a large measure of the self-belief that went with it. Lacking a suitable institutional mechanism through which to express its collective security concerns, ASEAN turned to its Post-Ministerial Conference (PMC). Originally conceived as a forum for the discussion of economic issues, the PMC turned increasingly towards security matters following the reunification of Vietnam in 1975.[23] It has been credited as a key force in securing the withdrawal of Vietnam from Cambodia,[24] although it is not at all clear whether this success was due to the nature of the PMC itself, its informal atmosphere and the confidentiality of the discussions that take place there (that is, its adherence to the 'way'), or to the access it afforded ASEAN foreign ministers to those truly holding the reins of power. That the Paris Accords were brokered largely by non-Asian powers responding to global imperatives lends credence to the latter suggestion, although this has not prevented the PMC being used as the blueprint from which the ARF was constructed.

If the strategic landscape of the region was changing shape in the decade of the 1970s so, too, was its economic counterpart, and even more fundamentally. The regional economic context continued to be dominated by the Japanese presence, albeit complicated by a new sense of disjunction between Japan and the US made concrete by the launch of Japan's comprehensive security policy following the Nixon dollar devaluation and the later oil 'shocks', and by a growing sense of resentment within the region at the highly visible profile of Japanese business.[25] This major diplomatic initiative, however, which presupposed the successful implementation of a relatively more active and sensitive role for Japan in the diplomatic arena, largely trailed thinking within Japanese business circles which, in the face of the oil price rises and mounting domestic costs associated with steady wage increases (and, later, the growing power of the environmental lobby), had begun the process of moving their operations offshore, with Southeast Asia the chief destination. In these circumstances, a growing concern with the consequences for the ASEAN subregion as it slowly came to resemble a preserve of public and private interests located within Japan is entirely understandable. Embracing and rejecting aspects of Japanese economic and social power has been an almost constant theme in the relationship between ASEAN and its powerful neighbour, and hence in the articulation of the ASEAN way. Therein lies one source of the discord between promotion of a Japanese and an alternate Chinese heritage. This distinction is less important to our present purposes, however, than is an understanding of how the ASEAN way has been used as a defence against the metaphorical tidal wave of economic globalization that, in the eyes of some, threatens not only to blur the boundaries separating states and regions but to erase them completely.

ECONOMICS AND THE ASEAN WAY

As we have already suggested, the ending of US hegemony foreshadowed a degree of unilateralism in US foreign policy perhaps not seen before. This reached its heights in the Reagan era most noticeably with the prosecution of a 'Second' Cold War against the Soviet Union, the genesis and conclusion of which is well understood and relatively uncontested.[26] Less well understood and very much more contested are the effects of this new unilateralism in the economic sphere and the nature of its relationship to the globalization and/or regionalization of economic activity.[27] Rather than engage in detail with these arguments here, our present purpose requires us simply to acknowledge that, from one perspective, the policies pursued unilaterally by the United States have been both a (partial) cause and consequence of a process, beginning roughly in the early 1970s, that has seen the place and role of the state in the industrial core of the global economy – North America, Europe and Japan – slowly change, such that the classic formulation of the welfare or 'embedded' liberal state has either all but disappeared or been badly eroded. This new role – which Japan in particular appears to have resisted but which is now also being forced upon states outside the core by international organizations such as the International Monetary Fund (witness the Fund's recent insistence on the 'restructuring' of the South Korean and Indonesian economies) and the World Bank, as well as by the World Trade Organization and the APEC – is apparently for the state to act as facilitator for the free and untrammelled operation of market forces everywhere. In other words, the state's role is no longer to act as a buffer between the domestic economy and its international or global counterpart but as a 'transmission belt' for abrasive global market forces.[28]

As noted earlier, these abrasive global market forces contain the potential to blur the distinction between national boundaries and hence social formations. (Very few analysts, if any, would go so far as to suggest that such boundaries are destined to disappear completely.) In other words, even if states survive as dominant actors within the global political economy, they will come to resemble each other so closely in form and function that differentiation between them will be all but impossible. The reality may, of course, turn out to be very different, but even the potential loss of identity associated with globalization is enough of a threat to make this issue of central political import, and especially so in the post-colonial, economically dependent and vulnerable ASEAN states, the majority of whom have yet to achieve stability let alone harmony within their respective social milieu. There is a hierarchy here, in which the states nearest the core appear secure, while those further away grow progressively less so. ASEAN is far enough away from the core to feel constantly threatened by this potential loss of identity but near enough to feel capable of mounting a counter attack: one that not only seeks to improve its position in the hierarchy but to change the rules by which that hierarchy is configured. This is the multi-levelled task of the ASEAN way in the economic sphere (and also in the political sphere, as we shall see below). Unfortunately, improving the immediate economic position of ASEAN within the global economy can only be achieved by engaging fully with that economy and thus leaving its collective (or separate) social formation(s) open to further, faster erosion; whilst changing the rules implies confronting those who set those rules in the first place and gain the most benefit from them. In both cases this means challenging the economic and/or social power of the United States. The cumulative difficulties implied by the interactions of these multiple dilemmas gives us an indication of the likelihood of success, and thus goes some way to explaining the bitterness of those in the region who condemn the political pressure exerted by the US in particular for a 'level' economic playing field.

Nevertheless, the attempt is being made. The AFTA, to which we will turn in a moment, represents a major effort by the ASEAN states to improve their position within the hierarchy, while arguments that the origins and therefore strengths of the East Asian social formation differ from that of Europe and North America so fundamentally that it has either had the capacity to resist the forces of globalization to a far greater extent than other regions or that East Asian states have reacted to them in very different ways constitute the major thrust of the attempt to change the rules which govern that hierarchy.[29] In response to these latter arguments, one may simply say that although they do have some force they are frequently overdrawn, choosing to ignore the colonial legacy and five decades of Japanese influence outlined above, both of which, despite their negative social impacts, have also had beneficial (that is, growth-enhancing) economic effects that should be neither dismissed nor ignored. To attribute the recent economic success of ASEAN solely to the superiority of the East Asian social formation is, then, to ignore a wider historical context provided most importantly by the exigencies of the Cold War conflict. Moreover, these arguments consistently downplay the fact that all the ASEAN states are fundamentally dependent on trade and investment with and from the United States, Japan and Europe and are therefore inextricably tied into the global economy.

As for what are tantamount to racist claims that East Asia forms a coherent whole by virtue of a shared Confucian heritage, one only has to remind oneself that Imperial Japan attempted the same tactic during the

Pacific War, and failed. There is no single 'model', Japanese, Confucian or otherwise, upon which other East Asian states may build a defence against the forces of the global economy, as recent events have made all too clear.[30] There may well be a common thread of corruption and graft running through most if not all of the countries thus far affected by the Asian crisis, but a thread is all it is and one far too thin to allow us to bundle the Asian countries together into a coherent model of political economy. If, then, ASEAN has not as yet been able to change the rules by which the hierarchy is governed and seems unlikely to do so in the foreseeable future, it remains for us to explore how far it has succeeded in changing its position within the hierarchy. That, we recall, was the central function of the AFTA, and here too we find only failure.

AFTA and 'open' regionalism

For the first twenty years and more of its existence, effective formal economic cooperation between the member states of ASEAN was virtually nil. The ASEAN Concord of 1976 did provide for cooperation in trade and industrial development, banking and finance, food and energy, transport and tourism and a common external economic policy but failed to bring about any substantive increases in either intra-regional trade or industrial cooperation, although there was a little more success in this latter area. Likewise, the ASEAN Preferential Trading Arrangement was launched in the following year.[31] In part, this relative failure may again be attributed to events in the strategic arena, with the ASEAN Concord seen more as a rhetorical device through which to foster the appearance of cohesion between its members following the communization of Indochina rather than as a set of concrete proposals for shared economic development. Be that as it may, a basic lack of complementarity amongst the ASEAN economies, combined with and reinforced by continued dependence on the economies of former colonial powers, were central factors contributing to the relative failure of these initiatives.

It was not (sub)regional but individual initiatives undertaken during the 1980s that rescued the ASEAN states, as Indonesia, Malaysia, the Philippines and Thailand each began to turn away from import-substitution and towards the path of export-oriented economic growth travelled so successfully before them by Japan, South Korea and Taiwan. Uneven though the success of this movement has been, the spectacular rates of growth, raised levels of industrial competence, competitiveness and complementarity that followed served to draw the economic interests of all ASEAN members more closely together and fostered the 'domestic' conditions upon which the AFTA could be built.[32] By the early 1990s this congruity, in combination with the visible fallout of the globalization process – that is, the widening and deepening hegemony of the neo-liberal orthodoxy (expressed clearly in the creation of APEC); heightened fears of exclusive regional blocs centred around 'Fortress' Europe and the North American Free Trade Agreement; the liberalization of the Chinese economy, itself competing for scarce financial capital – and the lower strategic salience of the ASEAN subregion after the Cold War served to create a political and economic climate in which the logic of an AFTA could not be denied.

AFTA was duly ushered into existence in 1992. With a potential market of around 500 million (if Vietnam is included), a combined purchasing power of more than US$400 billion, a wide range of resource and skill endowments and the rapid speed of the proposed elimination of tariff and non-tariff barriers by 2003, the economic future looked set to live up to the flowery rhetoric accompanying its launch. As recent events have shown, the reality has turned out to be somewhat different; and while it is tempting to lay the blame for this on mismanagement, or corruption or even something as apparently mundane as the several problems associated with tariff harmonization, not the least of which is the absence from the AFTA framework of a dispute-settlement mechanism, the underlying problem is the fundamentally weak position of the ASEAN economies.[33] The true purpose of the AFTA, then, is 'to improve the grouping's competitiveness as a production base and investment location' in an era characterized by harsh and growing economic competition amongst all states, 'communist' China included.[34] In other words, the aim is to subject ASEAN (and by implication, Southeast Asia entire) to the full discipline of the market in the hope that low costs associated with cheap labour and limited regulatory standards coupled with an unevenly but increasingly skilled workforce accustomed to providing for their own welfare and disciplined to the realities of global competition will attract sufficient investment to keep the 'miracle' going.

As we have argued repeatedly, the future does not bode well. Moreover, questions remain as to the form and function of the AFTA, especially in the context of APEC to which all the established ASEAN states belong and which espouses the same broad goals. Why is AFTA so determinedly portrayed as a form of open and/or soft regionalism? How and why does it differentiate itself from APEC? We suggested one possible answer above: that AFTA has ideological as well as simply economic goals in that it seeks discursively to contrast itself with more 'exclusive' forms of regionalism. Putting what is largely the rhetoric of the

ASEAN way aside, however, a more convincing answer to the degree of openness of the AFTA lies again in the fact of continued dependence on trade and investment with countries and regions outside Southeast Asia. No amount of culturally specific individual or group 'resilience' can alter this reality. Rather than improving its position in the hierarchy then, soft, open regionalism has become the mask ASEAN has chosen to wear in order to disguise the ugly truth of this weakness and to obscure the unmentioned benefit of arguments surrounding the ASEAN way: that nothing is said about the losers in this equation, those unfortunates – women, the unemployed and unemployable, the ethnically disenfranchised and the rural poor – who disappear beneath the rhetorical rubble.

POLITICS AND THE ASEAN WAY

If the ASEAN way masks some of the weaknesses and faults inherent in the socio-economic circumstances obtaining within ASEAN then it has also been used to the same effect in the political and strategic spheres. In what are to varying degrees culturally and racially heterogeneous societies, the ASEAN way has for example been deployed as a unifying political strategy in a social context often characterized by blatant racial discrimination – with Malaysia, Singapore and Indonesia being the usual suspects.[35] However, there is one area, the resolution and prevention of conflict, in which the attempted application of the ASEAN way, although daring, seems more appropriate, even laudable. Before we turn to the key institutional nexus – the ARF – within which these attempts have coalesced in the 1990s, it may prove useful to return briefly to the topic of past efforts in this direction. What we will find is that these efforts failed not because they did (or could) not partake of the ASEAN way but for concrete historical reasons that ASEAN was either not subject to or managed to circumvent as part of a learning experience.

As we noted earlier, there have been a number of previous attempts at building multilateral alliances in Southeast Asia. Chief amongst these was SEATO, a US-sponsored organization inclusive of Asian and non-Asian members (the US, Britain, France, Australia, New Zealand, Pakistan, Thailand and the Philippines) which was itself matched by a Soviet proposal for an Asian Collective Security System. The latter initiative was still-born, while the former never took hold and was ended in 1977 – both suffering firstly from the fears and suspicions of the smaller (Asian) states that they would be dragged into the superpower conflict, and secondly from the resentment engendered by the appropriation by non-Asian powers of Asian status. In contrast, the Anglo-Malayan Defence Agreement which became the consultative Five-Power Defence Arrangements following the British withdrawal from east of Suez has survived to the present, although its very status as a 'consultative' and therefore informal arrangement speaks volumes as to why this may be so. Moreover, it has also served an at times much needed role as a facilitator to confidence-building between its two Asian members, Malaysia and Singapore.[36] Nevertheless, the Five-Power Defence Arrangements have never been invoked and are unlikely ever to be so, tainted as they are by the faint but lingering whiff of colonialism.

The ASA and Maphilindo, the indigenous attempts at multilateralism within the region, suffered from a mixture of the Cold War-related problems associated with those attempts outlined above and intra-regional rivalries reflective of both power-political and racial aspirations. Established in 1961, the pro-western ASA stood no chance of success in the face of the hostility of the Sukarno regime in Indonesia, which by then had moved somewhat away from a policy of aggrandizement through non-alignment and leadership of the so-called Newly Emerging Forces, towards aggrandizement through reliance on support from the Soviet Union and China. Likewise Maphilindo, which, as an organization established along racial lines, also stood no real chance of success in the face of Chinese opposition potentially emerging from the large presence of overseas Chinese in the region and (presumably) from Singapore after its establishment in 1965. In any case, the organization disappeared beneath Sukarno's 'Crush Malaysia' campaign launched in 1963.

This leaves us with bilateral defence cooperation within the region, the success of which, somewhat ironically, can in part also be traced to the confrontation between Malaysia and Indonesia. The cessation of the conflict in 1966 may not have ended rival claims over territory between ASEAN states (as the clash between the Philippines and Malaysia over Sabah was to prove), or the support of separatist movements by rival states (for example, Malaysian support for the Moro National Liberation Front in the Philippines); but what it *did* do was shock the non-communist states of the region into the realization that they shared a set of interests and goals chief amongst which were the eradication of communist subversion and the pacification of the various insurgencies entrenched within their domestic environments. Indeed, so numerous have these arrangements become that we cannot hope to delve into individual cases here:[37] suffice it to say that such problems have been dealt with bilaterally because either the threat perception has been shared (that is, with communist groups), or because the porous nature of the borders separating these states have allowed such groups to base themselves outside the territory in which they operate (for example, Malayan Communist Party guerrillas in the Thai–Malaysian border area) and/or their operations straddle established borders. In each case cooperation has been seen to be more effective than individual efforts at resolution. Nevertheless, serious problems between ASEAN states have yet to be resolved, chiefly in the area of disputes over territorial boundaries on land and at sea. Far from being resolved in the post-Cold War era these disputes look to be in danger of exacerbation as issues of fishing and, particularly, mineral rights creep inexorably to the top of the international agenda. It may be true that the ASEAN way has thus far prevented these disputes from flaring into open confrontation, but it is equally the case that it has failed to bring them to a close. This failure explains in large part why close bilateral cooperation between ASEAN states has not blossomed into a multilateral organization or defence community, whilst at the same time raising serious doubts as to whether the ASEAN way can continue to prevent open confrontation in the future: particularly in light of evidence at the very least suggestive of the emergence of competitive arms-racing in the region.[38]

It is tempting to argue that the realization that the ASEAN way was no longer sufficient to contain intra-ASEAN rivalries played a major part in the decision to push forward with a formal organization on the lines of the ARF. This does not explain, however, the fact that the scope of the ARF stretches far beyond the territory encompassed by ASEAN and indeed beyond Southeast Asia as a whole. It also highlights the defensive, negative aspects of ASEAN thought processes at the expense of ignoring the positive ideological potential inherent within the ASEAN way: a potential that opens the way for the ARF to become the institutional embodiment of arguments (also deployed, we recall, in reaction to the forces of globalization and in opposition to US-sponsored attempts at enforcing universal acceptance of the discipline of the free market) suggestive of the unique attributes and qualities of the East Asian social formation. Once again, however, we find that these efforts have as yet come to little or nothing.

The ARF: ASEAN writ large?

The ARF, which was formally brought into being by ASEAN and held its first meeting in July 1994, has from the start been hailed as an informal venue in which disparate countries of the Asia Pacific could meet in order to discuss security issues of mutual concern. Perhaps the most important decision taken by the foreign ministers of the 18 countries in attendance at the first meeting (ASEAN plus Australia, Canada, the EU, Japan, New Zealand, Republic of

Korea, the US, China, Russia, Laos, Papua New Guinea and Vietnam) was to reconvene on an annual basis. The second meeting proved more substantive, and resulted in the setting up of three joint working seminars each co-chaired by an ASEAN and non-ASEAN country: one on peacekeeping operations (Malaysia and Canada), one on search and rescue missions (Singapore and the US), and one on confidence-building measures (Indonesia and Japan). In addition, the ARF process was recognized to be operating on a 'two track' basis, with government-led discussions informed by and reacting to input from strategic institutes and other non-governmental organizations based in member states, which in turn took their cue from decisions reached at the formal level. The third meeting, held in Jakarta on 23 July 1996, added a fourth working group on disaster relief (co-chaired by Thailand and New Zealand) and discussed limiting the 'geographic footprint' of the ARF to North and Southeast Asia and Oceania as well as setting out criteria for the entry of new participants, and drawing attention to the signing of the Southeast Asia Nuclear Weapons Free Zone (SEANWFZ) Treaty in December 1995. The fourth meeting, held in Subang Jaya, Malaysia on 27 July 1997 continued in very much the same vein, reporting on the activities of the working groups on both 'tracks' and discussing, *inter alia*, the situation in the South China Sea, Cambodia and North Korea as well as the contribution that economic growth and cooperation had made to the continuing peace and stability of the region. This latter point notwithstanding, there was also a recognition that the stability of the region rested in large measure on the strength of relations between the major powers – China, Japan, the Russian Federation and the United States.[39]

Given the sudden appearance of the ARF and its success since its inauguration (with 'success' initially being gauged by the fact that its life extended beyond the first meeting) apparently against all the historical odds, it is no surprise that scholarly inquiry has sought to clarify exactly how and why the ARF process began when it did. The standard explanation, which – with the addition of the comments pertaining to the strength of the ideological component made above – for the most part we accept here, proceeds as follows. In the context of the end of the Cold War the US, aware of its relative decline, was able and sorely tempted to turn inwards in order to remedy those domestic faults thought to be responsible, potentially leaving a 'power vacuum' in East Asia which by definition had to be filled by one or both of the two remaining states with the capacity so to do in the foreseeable future, China and Japan. Forewarned by the historical record of the consequences of the mutual schizophrenia attending the relationship between these two countries, all states (including China and Japan) claiming membership of the Asia

Pacific community had a vested interest in addressing this potentially explosive situation. Seeing no credible alternative the US was persuaded to remain engaged, although in a more limited role. The ASEAN states, concerned with their own subregional set of security dilemmas as well as with these wider issues, and slightly paranoid about the prospect of being left to cool their collective heels in a relatively insignificant strategic backwater, stepped into the institutional breach proffering as their credentials their unique approach to conflict-prevention and confidence-building and almost thirty years' experience.

This is all well and good, but it has not yet adequately explained why ASEAN played the role of formal initiator and guiding light rather than one or a combination of the major powers as suggested by both practice and theory.[40] Two answers seem possible, only one of which also appears likely. The first is again that the historical circumstances attending the end of the Cold War favoured an ASEAN-led organization over any other; the second, that the complexity of the security issues and problems faced in the East Asian region in tandem with what for Westerners is effectively an alien social formation militated against Western approaches to negotiation and problem-solving (which are characterized by formality, legality, hostility and so on) and called instead for an approach indigenous to the region: that is, the ASEAN way. Thus ASEAN undertook 'the obligation to be the primary driving force' in discussions of regional security because of its unique cultural background, its successful historical record based upon adherence to this 'way' and the unsuitability of other nominees from both the West and East.[41]

The second of these alternatives is unconvincing. One does not even have to deny the existence or strength of the ASEAN way in order to puncture it, since the weight of evidence is overwhelmingly on the side of the first of the answers suggested above: ASEAN was allowed and even encouraged to set up and run with the ARF because it suited the requirements of the major powers. That it also allowed ASEAN a modicum of pride in its Asian heritage and a feeling of security in an uncertain world was a welcome but not critical extra. This is not to deny the potential contribution that the ASEAN way may make to the ARF process, but simply to point out that for the leading states the calculations pertaining to the benefits of an ASEAN-led organization were not based primarily on the degree of likelihood that the ASEAN way would make an important difference. Rather, they were based upon sets of criteria that differed from state to state.

For the US, the decision to participate had strategic, economic and ideological dimensions. In the strategic arena, participation would serve to reassure East Asian states of continued US commitment despite the reduction of US troop levels following the implementation of the 1990 East Asia Strategy Initiative and the withdrawal from the Philippines. Meanwhile, the increased salience of economic issues and relations after the Cold War threw into stark relief the series of trade disputes between the US and its Asian allies, some of which were beginning to threaten what had until then been fairly stable strategic relationships. Hence, the economic and strategic interests of the US have been brought together and linked to those of its allies through the continuous propagation of a neo-liberal ideology demanding, at the surface level, of economic conformity but ultimately and at a deeper level of the atomization of society into a fragmented terrain populated by individual consumers estranged from one another and from the community at large. The rhetorical separation of politics from economics and the pursuit of 'free' trade serves these interests and aims perfectly, although it has at one and the same time served to raise the hackles of its Asian allies. Engagement through multilateral fora such as the APEC and ARF enables the pursuit of these interests and thus constitutes a further weapon in the armoury of US foreign policy rather than the first step towards the abandonment of the unilateral and/or bilateral approaches that Washington has preferred in the past.

For Japan, the decision to participate in the ARF came partly from a sense that it was time for it to play a greater role in strategic affairs commensurate with its status as an economic superpower, and partly as a response to the swingeing criticism it suffered from certain sections of the international community over its contribution to operation Desert Storm. Uppermost in Japanese minds, however, was the need to keep the US engaged in the region and particularly to ensure the perpetuation of its own bilateral security relationship, the abrogation of which would almost certainly require Japan to revise its security strategy and possibly even its 'Peace' Constitution. Given the historical record and the extent of its economic involvement in Southeast Asia the choice of location must have seemed almost preordained, although an additional incentive was the lack of an alternative geographic or institutional venue. To have located it in Northeast Asia would have been to exclude the Southeast Asian countries, while attaching it to the APEC would have alienated Malaysia in particular. For China the decision to participate reflects its concern that the US and Japan were attempting to contain and isolate it, as the US and Soviet Union had done in the past. The ARF gives China, whose territorial claims in the South China Sea have effectively bridged the former gap between North and Southeast Asia, a forum in which to express its views and in which it can keep an eye on the activities of its rivals.

This leaves us with ASEAN. Why did it put itself forward as the 'primary driving force' behind the ARF and what benefits does it expect to gain? To keep itself safe and relevant is the simple answer, although 'safety' and 'relevance' are in fact two sides of the same coin. Participation in the ARF gives ASEAN an additional weapon with which to confront its traditional problems (although these will inevitably continue to be thrashed out behind the veil of ASEAN confidentiality) if it so wishes, and the recently arrived issues such as the environment, oil and so on. It also allows ASEAN additional reassurance in the face of the arrival of China in Southeast Asia, and a place at the 'top table' of international diplomacy. At a different level, however, it also provides ASEAN with a venue in which it can push forward with its own ideological agenda. For ASEAN, western approaches to conflict resolution and prevention do not work; or work only at enormous monetary, social and political cost. What is more, these approaches have on several occasions within the past fifty years led the world to the brink of extinction and back. Thus to allow a western-led organization to come into being, or an organization run along western lines and precepts but led by a coalition of Asian states would not in any way enhance ASEAN security but would in fact threaten it. Attempts to apply the ASEAN way to the security dynamic obtaining in East Asia and the Asia Pacific are not, then, simply or only political rhetoric deployed in order to mask socio-economic weakness and/or strategic fragility, but a genuine expression of concern over the inadequacies of western solutions and forthright confidence in Asian values as the foundation upon which ASEAN-led approaches have achieved success.

There is, however, very little hope that the ASEAN way will for long be the model upon which the ARF is constructed, if indeed that has ever been the case. From the outset some of the non-ASEAN members, the Republic of Korea in particular but also Australia and the US, have been unhappy with the appropriation of the leadership of the organization by ASEAN, with some pushing quietly for it to take the form of an 'Asian' Regional Forum. Similarly, the 'evolutionary,' consensus-based approach adopted by the ARF process is proving increasingly irksome to some of the participants, although said approach was again endorsed at the 1997 meeting. Chinese sensitivities are a major area of concern, with the issue of Taiwan ruled off the agenda completely. Such are the niceties that have to be observed with respect to the Chinese that the inter-sessional seminar on confidence-building measures is referred to as a 'Group' not a

'Meeting' as is the case with the other seminars so as to give the impression that it would function only on an *ad hoc* basis.[42] Add in the problems associated with the absorption of new members into ASEAN, none of which are even familiar with the organization's collegiate style let alone steeped in it; new issues such as protection of the environment; old issues but new faces such as the divided Korean peninsula; and seemingly intractable contradictions between cherished ASEAN goals such as the SEANWFZ and its ultimate reliance on the nuclear protection offered by the US, and it appears less and less likely that the ARF will fulfil the dream role of an ASEAN writ large.

CONCLUSION

The account offered of the causal connections between the end of US hegemony and the rise to prominence of the ASEAN way has painted a picture of that 'way' that is bleak, sceptical and pessimistic, preferring to locate the source of the cohesion of the ASEAN states not in shared racial/cultural attributes but in the unique historical circumstances obtaining in the early post-War years. From this perspective the ASEAN way appears as a last line of defence against the economic and social power of the advanced industrial states of the West, an ideological redoubt hastily thrown up in reaction to the material and ideological onslaught launched by the US in the 1970s which has gradually but exponentially grown in strength and effect. The AFTA and involvement in the ARF have thus far failed to improve the position of ASEAN within the hierarchy of states, while the ASEAN way has largely failed to deliver a change in the rules by which that hierarchy is configured. There is no doubt, however, that through the mechanism of the ARF the ASEAN way has and may yet make a contribution to the prevention and resolution of conflict in East Asia and in so doing will show that the ideas and values with which it is imbued can be of great value.

In the economic arena, by contrast, the case for an ASEAN way seems more tenuous. The economic crisis currently affecting not only some of the key ASEAN states but also the Republic of Korea and Japan weakens if not destroys claims to uniqueness based on shared cultural and/or racial affinities and attributes. It will, one hopes, also finally put paid to claims that there is a discernibly Asian model of socio-economic development. Far too much effort has been poured into the search for this Asian Holy Grail, and far too little into uncovering and reporting the flip side of the

East Asian 'miracle' – the poverty, injustice and corruption to which we have here been able only to allude. The great tragedy of the ASEAN way is that in seeking to defend the status quo its proponents simultaneously perpetuate and reinforce inequalities and injustices that not only blacken the social fabric but threaten to ignite it. With this in mind, it seems more than a little ironic that media and diplomatic attention is beginning to focus upon the exacerbation of the trade disputes that inevitably follow currency realignments associated with such crises when they should perhaps be drawing attention to the exacerbation of the hardship and trials suffered by those producing the goods concerned.

NOTES

1. Cited in M. Antolik, *ASEAN and the Diplomacy of Accommodation* (New York: M.E. Sharpe, 1990), p. 6. The relevant documents are currently available at ASEAN's official website at http://www.asean.or.id/.
2. M. Leifer, *The ASEAN Regional Forum*, Adelphi Paper 302 (London: Brassey's for the IISS, 1996), p. 11.
3. D. SarDesai, *Southeast Asia: Past and Present* (Boulder, CO: Westview Press, 1997), 4th edn; G. Segal, *Rethinking the Pacific* (Oxford: Oxford University Press, 1990).
4. M. Yahuda, *The International Politics of the Asia-Pacific* (London: Routledge, 1996).
5. A. Acharya, *A New Regional Order in South-East Asia: ASEAN in the Post-Cold War Era*, Adelphi Paper 279 (London: Brassey's for the IISS, 1993).
6. D. Emmerson, 'Region and Recalcitrance: Rethinking Democracy through Southeast Asia', *The Pacific Review*, 8, 2 (1995), pp. 223–48; J. Wanandi, 'ASEAN's Domestic Political Developments and their Impact on Foreign Policy', *The Pacific Review*, 8, 3 (1993), pp. 440–58. R. Robison and D. Goodman (eds), *The New Rich in Asia: Mobile Phones, MacDonalds and Middle-Class Revolution* (London: Routledge, 1996).
7. R. Higgott and R. Stubbs, 'Competing Conceptions of Economic Regionalism: APEC versus EAEC in the Asia Pacific', *Review of International Political Economy*, 2, 3 (1996), pp. 516–35. See also Chapter 10.
8. World Bank, *The East Asian Miracle: Economic Growth and Public Policy* (Oxford: Oxford University Press for the World Bank, 1993).
9. We return to this below, but for more historical detail see Antolik, op. cit.
10. A. Acharya, 'Ideas, Identity and Institution-building: from the "ASEAN Way" to the "Asia-Pacific Way"', *The Pacific Review*, 10, 3 (1997), pp. 319–46 (pp. 328–33).
11. Hereafter we will employ only the term 'ASEAN way'. Y. Funabashi, 'The Asianization of Asia', *Foreign Affairs*, 72, 5 (1993), pp. 75–85; Mahbubani Kishore, 'The Pacific Way', *Foreign Affairs*, 74, 1 (1995), pp. 100–11; and Chin Kin Wah, 'ASEAN: Consolidation and Institutional Change', *The Pacific Review*, 8, 3 (1995), pp. 424–39.

12. According to our taxonomy the ARF is also a meta-regional organization but, significantly, one formally proposed by ASEAN itself. For a theoretical discussion of the emergence of the 'new' regionalism, see A. Hurrell, 'Explaining the Resurgence of Regionalism in World Politics', *Review of International Studies*, 21, 4 (1995), pp. 331–58.
13. S. Yamakage, 'Human Rights Issues in Southeast Asia', *Japan Review of International Affairs*, 11, 2 (1997), pp. 118–37; D. Mauzy, 'The Human Rights and "Asian Values" Debate in Southeast Asia: Trying to Clarify the Key Issues', *The Pacific Review*, 10, 2 (1997), pp. 210–36.
14. For a similar attempt see M. Bernard, 'Regions in the Global Political Economy: Beyond the Local-Global Divide in the Formation of the Eastern Asian Region', *New Political Economy*, 1, 3 (1996), pp. 335–54.
15. For an introduction to this line of argument see R. Stubbs, 'The Political Economy of the Asia-Pacific Region', in R. Stubbs and G. Underhill (eds), *Political Economy and the Changing Global Order* (London: Macmillan, 1994), pp. 366–77.
16. J. Dower, 'The Useful War', in J. Dower, *Japan in War and Peace: Essays on History, Culture and Race* (London: Fontana, 1996), pp. 9–32.
17. M. Bernard, 'Post-Fordism, Transnational Production, and the Changing Global Political Economy', in Stubbs and Underhill (eds), op. cit., pp. 216–29 (pp. 218–21); J. Moore, 'Democracy and Capitalism in Postwar Japan', in J. Moore (ed.), *The Other Japan: Conflict, Compromise, and Resistance since 1945* (Armonk, NY: M.E. Sharpe, 1997), pp. 353–93.
18. Japan also, of course, became somewhat more than just a 'second rank' economy.
19. See R. Steven, 'Japanese Investment in Thailand, Indonesia and Malaysia: A Decade of JASEAN', in Moore (ed.), op. cit., pp. 199–245.
20. The literature is vast. For a selection of recent views see the following: J. Fallows, *Looking at the Sun: The Rise of the New East Asian Economic and Political System* (New York: Vintage Books, 1995); W. Hatch and K. Yamamura, *Asia in Japan's Embrace: Building a Regional Production Alliance* (Cambridge: Cambridge University Press, 1996); and P. Katzenstein and T. Shiraishi (eds), *Network Power: Japan and Asia* (Ithaca: Cornell University Press, 1997).
21. M. Berger, 'The Triumph of the East? The East Asian Miracle and post-Cold War Capitalism', in M. Berger and D. Borer (eds), *The Rise of East Asia: Critical Visions of the Pacific Century* (London: Routledge, 1997), pp. 260–87 (p. 268).
22. SarDesai, op. cit., p. 301.
23. In the 1990s the PMC also turned towards issues of ideology: it was at the 1993 meeting of this body that the EAEC was launched.
24. Yuen Foong Khong, 'ASEAN's Post-Ministerial Conference and Regional Forum: A Convergence of Post-Cold War Security Strategies', in P. Gourevitch et al. (eds), *United States–Japan Relations and International Institutions after the Cold War* (San Diego: University of California Press, 1995), pp. 37–58.
25. T. Akaha, 'Japan's Comprehensive Security Policy', *Asian Survey*, 31, 4 (1991), pp. 324–40; S. Sudo, *The Fukuda Doctrine and ASEAN: New Dimensions in Japanese Foreign Policy* (Singapore: Institute of Southeast Asian Studies, 1992).

26. F. Halliday, *The Making of the Second Cold War* (London: Verso, 1986), 2nd edn.
27. J. Zysman, 'The Myth of a "Global Economy": Enduring National Foundations and Emerging Regional Realities', *New Political Economy*, 1, 2 (1996), pp. 157–84; J. Perraton *et al.*, 'The Globalisation of Economic Activity', *New Political Economy*, 2, 2 (1997), pp. 257–77; S. Gill, *American Hegemony and the Trilateral Commission* (Cambridge: Cambridge University Press, 1990), pp. 105–11.
28. R. Cox, 'Global *Perestroika*', in R. Miliband and L. Panitch (eds), *The Socialist Register* (London: Merlin, 1992), reprinted in R. Cox with T. Sinclair, *Approaches to World Order* (Cambridge: Cambridge University Press, 1996), pp. 296–313.
29. There is again some overlap since the discursive representation of AFTA as a form of 'soft' or 'open' regionalism does not simply or only fit in with the rhetorical commitment to free trade espoused by meta-regional organizations such as APEC, but does so in a way that draws attention to the comfort of that 'fit': the implication being that ASEAN regionalism in whatever form could *never* be exclusionary. For a discussion of the concept of 'soft' and/or 'open' regionalism as espoused by ASEAN see Acharya, 'Ideas, Identity and Institution-building', op. cit.
30. At the time of writing the quality press is saturated with news of the economic crisis rolling over East Asia: see the *Financial Times* throughout the months of November and December 1997 and January 1998.
31. For details see Siow Yue Chia, 'The Deepening and Widening of ASEAN', *Journal of the Asia Pacific Economy*, 1, 1 (1996), pp. 59–78 (pp. 60–65).
32. See P. Bowles and B. Maclean, 'Understanding Trade Bloc Formation: The Case of the ASEAN Free Trade Area', *Review of International Political Economy*, 3, 2 (1996), pp. 319–48.
33. For details on tariff harmonization, see Chia, ibid., pp 66–70; Tsao Yuan Lee, 'The Asean Free Trade Area: The Search for a Common Prosperity', in R Granaut and P. Drysdale (eds), *Asia-Pacific Regionalism: Readings in International Economic Relations* (Pymble, NSW: Harper Educational, 1994), pp. 319–26. By the year 2010 an Asian investment area is expected to be in place, and by 2020 the plan is to allow the free-flow of capital among member states.
34. The citation is from Chia, op. cit., p. 60
35. C. Berger, op. cit.
36. Leifer, op. cit., p. 10
37. For details see Acharya, 'A New Regional Order in South-East Asia', op. cit., pp. 27–30
38. Acharya, ibid., pp. 30–40; K. Calder, *Asia's Deadly Triangle: How Arms, Energy and Growth Threaten to Destabilize Asia Pacific* (London: Nicholas Breley, 1996).
39. See the respective Chairman's Statements issued following each meeting of the ARF (available at the website cited at note 1 above); see also Leifer, op. cit., for details of the first two meetings. Cambodia, India and Myanmar have joined in the interim, whilst applications from Britain and France for membership separate from the EU are believed to have triggered the debate over criteria for new admissions.

40. For a discussion see Yuen Foong Khoong, 'Making Bricks without Straw in the Asia Pacific?', *The Pacific Review*', 10, 2 (1997), pp. 289–300.
41. The citation is from the Chairman's Statement, *The Second ASEAN Regional Forum*, Brunei Darulsalam, 1 August 1995.
42. This last point and much of the preceding detail is from Leifer, op. cit.

9 Politics of Identities and the Making of the 'Greater China' Subregion in the Post-Cold War Era

Ngai-Ling Sum

The 'Greater China' subregion (that is, Hong Kong, Taiwan and Southern China) which provides my focus below differs from the other subregions analyzed in this collection. Some of the latter have a long history or, at least, an institutional and organizational infrastructure that gives them a certain path-dependent solidity and coherence. The 'Greater China' subregion, for all that it has an invented history provided by the People's Republic of China (PRC), is a recent product of largely bottom-up exchanges among economic and social actors from Hong Kong, Taiwan, and southern China. To the extent that it is now gaining institutional shape, it is in the form of network-like patterns which are partly influenced by path-shaping struggles around the politics of identity and the reimagination of the subregion.

These identity struggles and reimaginations are connected, in turn, to the strategic context constituted by global–regional–national changes. The latter comprise: the globalization and triadization of capitalism (linked to, but not reducible to, the crisis of Atlantic Fordism, and expansion of specific commodity chains); the end of the Cold War, which was marked by bipolarity in geopolitical and security terms, and the scope this has created for an increased importance of Japan and China as the current and emerging regional hegemons in a more multipolar world; and the contradictory rise of more cosmopolitan and 'tribal' identities alongside nationalist ones. These changes provide the contexts in which economic, social and political actors in the subregion pursue their competitive, security, and nationalist goals. This leads to a complex interplay of geo-economic, geopolitical, and nationalist discourses/strategies and creates the space for wide-ranging and often contradictory interests/identities for subregional social forces.

Thus this chapter argues that the making of the 'Greater China' subregion is dominated by geo-economic and nationalist discourses/

strategies. These are constituted and facilitated in and through a strategic network of transnational and translocal actors with an imagined community of interests. Moreover, given the socially embedded character of such networks, they are also discursively articulated with other identities/ interests related to other geometries of powers. For example, the geo-economic identity of a subregion may become a focus of intervention or interruption by competing global/regional hegemons. In the case of 'Greater China', for example, the subregion is being showcased both as a geopolitical hub for 'democracy' by the US/UK and as a 'nationalist' powerhouse by the PRC. In this regard, social forces in China, Hong Kong, and Taiwan are increasingly confronted with dilemma-full opportunities rooted in the interpenetration of the geo-economic/national and geopolitical time and space. The resulting conjuncture is characterized by a dialectical interplay of 'multiple consciousness' which marks this subregion as contested space. These clashes of discourses/practices come to be (in)fused with power and ideology and thereby engender what can be seen as politics of identities in the subregion. Such identity struggles not only mark the subregion as a contested space; they also mark it as a space of reimagination (for example, 'Greater China' as a high-tech subregion or 'Greater Shanghai' as the new centre of economic gravity for the twenty-first century). Let us start with the geo-economic discourses and the strategies of the PRC.

THE EMERGENCE OF THE DOMINANT GEO-ECONOMIC/ NATIONALIST DISCOURSES AND STRATEGIES IN THE 'GREATER CHINA' SUBREGION

The geo-economic discourses and strategies: the 'Open Door' project of the PRC

By the late 1970s, the Chinese leadership, especially Deng Xiaoping, had realized the need to construct a new hegemonic project that could offer a better living standard by 'opening the door' of China and to rejoin the world. The discourses and strategies of the 'open door' create opportunities for China to participate more fully in the global–regional–national nexus. Proposals for pioneering experimental sites have found resonance within the party elites and among key coastal-provincial actors in Guangdong and Fujian. These actors began to demand 'special/flexible measures' that would permit the creation of new geo-economic discourses and strategies that reconnect China to the global–regional system(s).

Politics of Identities Making of the 'Greater China'   199

The Party Central Committee responded with a decentralization strategy which enabled Guangdong and Fujian to adopt 'special policies and flexible measures' in 1979. This strategy privileged and consolidated a 5-year macro-contracted responsibility system which reformulated the central–local, and private–public relations in China. Here the commercial and fisco-financial reforms are most relevant. The former remaps central–local relations by allowing enterprises/business units under central control (except those in certain areas) to be managed by the province and to enter into contracts (subject to central approval) with incoming private investors of a value up to US$3 million (extended to US$10 million in 1985, US$30 million in 1988); and fisco-financial reform introduced in 1980 enables provinces such as Guangdong to retain the remaining surplus on foreign currency earnings from exports after paying 30 per cent of the export surplus to the central government.[1]

These reforms were applied to the coastal provinces of Guangdong and Fujian, especially in their Special Economic Zones (SEZs). Three such zones were located in Guangdong (Shenzhen, Zhuhai, and Shantou) and one in Fujian (Xiamen). These zones are earmarked for experimentation with the interface of global–regional–national–local economy. They are carriers of the following symbolic roles: (a) developing economic links between Hong Kong (Shenzhen), Macau (Zhuhai), Taiwan (Xiamen) and overseas Chinese communities (Shantou), with a view to reunifying China; (b) serving as 'windows' and 'pivots' for learning advanced technology and managerial skills as well as attracting foreign investment; and (c) serving as a 'test tube' or 'laboratory' for experimenting with economic reform at the intersection of socialism and capitalism. Since the establishment of the four SEZs in 1980, China has designated other open areas for foreign direct investment (FDI) for such interactions. These include: 14 coastal cities (such as Dalian, Shanghai, Tianjin, and so on) and three coastal areas (for example, Pearl River Delta in Guangdong). Foreign investment in these different types of open areas enjoys various incentives. These usually include an extensive package of tax incentives and exemptions from certain legislation and/or administrative orders that are not applicable to the rest of the economy. For example, the income tax rate of 15 per cent in SEZs was extended to coastal areas in 1992.

Given its headstart, Guangdong has the biggest concentration of open areas able to take advantage of FDI. In addition, the reform programme has offered Guangdong's provincial and local governments exclusive fiscal incentives to develop an outward-looking economy. According to a contract between Guangdong and the central government, the former was allowed to retain 70 per cent of export earnings above the 1978 export level (US$1.4 billion), while other provinces (except Fujian) could retain only 30 per cent for centrally managed goods and 40 per cent for locally managed goods. The provincial and local governments in Guangdong and Fujian thus had great incentives to actively enter into strategic partnership and subcontracting activities from Hong Kong/Taiwanese/overseas firms outsourcing for cheap labour, land and component parts for global and regional counterparts.

The nationalist discourse and strategy: the 'one country, two systems' formula

Given that the 'open door' policy has unleashed certain local forces, especially in the coastal provinces, Maoism can no longer be maintained as the sole narrative in defining the national identity of China. During the Maoist period, an ideology of anti-imperialist nationalism imagines an exploitative capitalist world as the enemy of an entire nation and threatens the independence of the Chinese people. However, the privileging of an 'open door' policy, which is pertinent to Deng's project, has provided the context for the reconsideration of China's national question. First, the 'open door' imperative, which imagines a mix of global interdependence and national competitiveness, conflicts with the 'enemy' narrative of Leninism; and, second, the privileging of the decentralization strategy contradicts the practices of a top-down/centralized party-state.

The decoupling of nationalism from (Leninist) anti-imperialism has created the discursive space for a (re)coupling of nationalism with geo-economic imperatives as a basis for seeking and organizing hegemony. In this regard, nationalism, which is coded in terms of the 'reunification' of territories with common 'pastness' (for example, common language, culture, and history), helps to homogenize an 'ethnically common' and 'economically-vibrant south'. This imagined community of 'Greater China' was concretized in and through the discourse of 'one country, two systems'. This operates in two interlinked ways.

First, the 'one country' rhetoric homogenizes the nationalist time/space of China, Hong Kong (Macau), and Taiwan as communal spaces with shared memories and culture (for example, 'descent from Dragon/Yellow Emperor/Beijing Man', 'love of the motherland') pertinent to ethno-national identity. Yet an essentialization of their commonality cannot ignore their embedded difference (for example, socialist China and capitalist Hong Kong and Taiwan). Thus, in this regard, the narrative of 'two systems' and the creation of the identity of 'special administrative regions (SARs) of China' allow for such differentiation and historical discontinuities within the communal space. The 'one country' and 'two systems' rhetoric is open to diverse interpretations. China prioritizes the former and codes it as part of its nationalist project; conversely, democrats in Hong Kong essentialize the latter and code it as the basis of their democracy movement in political space. Taiwan rejects this formula, however, because the people in Taiwan will not accept a subordinate position to Beijing. They prefer equal co-existence and the adoption of 'flexible diplomacy'.

Despite differences over this formula, China privileges the 'one country' strategy as a nationalist–cultural space linking China, Hong Kong (Macao), and Taiwan. In implementing this strategy, the PRC adopts a 'stick-and-carrot' approach. Beijing emphasized that it will not renounce the use of armed force, especially in relation to Taiwan, if the latter's independence-democracy movement thrives. On the other hand, Beijing offers incentives to attract Taiwanese and Hong Kong capital to invest in the 'motherland'. The 'carrot' side of the approach is important for its gradual coupling of geo-economics with nationalism. New economic-cultural codes such as the 'promotion of patriotic ethnic Chinese (*huaqiao*) to invest in the motherland' to form the 'Co-ordinating System of Chinese Economies'/'Southern China Growth Triangle' are used to remap a linguistically and culturally compatible space of Taiwan, Hong Kong, Macao, and Guangdong-Fujian provinces. This intermixing of the 'cultural' and 'economic' codes can be seen as a hybridized construction to form what the author elsewhere terms 'pragmatic nationalism'.[2] This hybrid helps in the reshaping of spatial complementarities in the subregion and in promoting its competitiveness in relation to global restructuring. In defining the identity of the subregion in this way, actors' perceptions and conduct are encouraged to cross borders within 'Greater China'. In particular, actors are urged to draw on their ethnic and cultural affinities within the subregion; and to engage in economic/social networking without directly addressing, at least in the short-term, the thorny problem of the relationship between nationalism and politics.

THE MAKING OF GEO-ECONOMIC/NATIONALIST CAPACITIES WITHIN THE 'GREATER CHINA' SUBREGION

In response to the above emerging discourses/opportunities within the subregion, the rise of Western protectionism, the US's granting of the Most Favoured Nation (MFN) trade status to mainland China, and increases in their own domestic costs for land and labour, Hong Kong's investors have searched for new production sites. It is now by far the biggest investor in Guangdong province, supplying 80 per cent of the FDI. The Pearl River Delta is now a major production base for its more labour-intensive products. It is estimated that almost 25,000 Hong Kong manufacturing enterprises, mostly in textiles and clothing, toys and consumer electronics, have moved there to exploit low labour and rent costs. They directly employ about three million workers, which is three times the total labour force of the manufacturing industries still operated in Hong Kong.

Unlike Hong Kong, the Taiwanese government in the early 1980s saw few advantages and many serious Cold-War risks in trade with the mainland. Commercial links were established in 1979 with trade between the two areas being conducted through Hong Kong. However, the appreciation of the Taiwanese currency from 1986, the rising cost of land, and the high standards set by the 1984 Basic Labour Law have all made the Chinese market increasingly attractive to Taiwanese businesses. The threat of investment strike by Taiwanese investors urged the government to replace restrictions on trade/investment by positive guidance by the Kuomingtang government. Such guidance for indirect trade with China began with the decision in November 1987 that Taiwanese citizens would be allowed to 'visit relatives' in China. The Central Bank of Taiwan also liberalized the foreign exchange regime in 1987 by allowing free capital outflows under five million new Taiwanese dollars. The legalization of travel and liberalization of foreign exchange sharply accelerated the growth in trade and FDI between them. In 1988, the Chinese State Council promulgated a set of 22 measures to encourage investment from Taiwan. Taiwan capital is treated as 'special domestic capital' and can engage in business not open to foreign capital, such as banking, wholesale and retail. In January 1993, the Taiwanese government began allowing Taiwanese companies to plough investments into China of US$1 million and below, without going through a third site. As a result, some five million Taiwanese have visited China and some 9300 Taiwanese firms have moved their production facilities there with investment of US$8.6 billion at the end of 1993. However, for security consideration, Taipei deems it necessary to restrict trade and movement of factors of production through a third site. In this sense, Taiwan's position on cross-Strait economic exchanges is to adopt a gradual and selective open-door policy while maintaining its competitive edge.

These reactions of the Hong Kong and Taiwan firms/organizations have consolidated strategic networks of actors who are involved in coordinating, mediating, and condensation of a geo-economic-nationalist formation. Of particular significance here is the consolidation of a network of networks that cross-cut central–provincial–local, public–private, and global–regional domains within China and the 'Greater China' subregion. Let us start with the emergence of strategic networks in China.

The emergence of central–provincial–local entrepreneurial networks in China

The 'open-door' policy has unleashed new central–provincial–local forces within China. These forces are mediated by a network of public, quasi-public and private institutions aimed at expanding foreign economic relations and development. *On the central level*, these include the Ministry of Foreign Economic Relations and Trade (MOFERT)[3] which manages the introduction of foreign investment, new trading arrangements, and establishment of joint ventures as well as the China International Trust and Investment Company (CITIC)[4] which was set up in order to facilitate the establishment of joint ventures. *On the provincial level*, Guangdong and Fujian followed the centre in setting up the Guangdong (Fujian) province functional and line ministries or state-owned companies to administer/negotiate/finance foreign trade. Further decentralization of the economy in 1984 accelerated this process when branches of foreign and industrial trading corporations under ministries set up provincial corporations. Guangdong and Fujian, in particular, were allowed to acquire many branches of the public-owned foreign trading corporations, and to set up their own organizations in Hong Kong and Macao for trade/investment promotion and information gathering. By 1987, the number of such corporations had risen to 1900.

These strategies/processes have also promoted changes *on the local level*. Here I will concentrate on the emergence of a strategic network in special localities, such as 'special economic zones' or 'development zones' on county–township–village level along the coastal provinces. On these levels, the *private–public network* includes: a) county party/government officials and the trading companies (*gongsi*) that they control (which must find foreign investors, export quotas, domestic markets, and foreign technology); and b) township–village–state/party officials and the township–village enterprises (TVEs) that they control. Often, the managers/executives of these new companies/enterprises are drawn from county-, township-, or village-level party and/or state functionaries.[5] Nee termed them 'cadre entrepreneurs'[6] and they can be seen as an intertwining relationship between private and public interests as well as central–local domains. For the most part, these 'cadre entrepreneurs' enjoy good connections to the higher state/party hierarchies.

These changes have led to the strong presence of a strategic network constituted by a set of new public actors/organizations at central, provincial, and local levels as well as a layer of quasi-public actors and enterprises encouraged to adopt a development strategy that is flexible, export- and profit-oriented. This network of private–public actors/organizations, especially those in growth- and export-oriented regions, are benefactors of the decentralization/coastal strategies. This involves the devolution of ownership rights (which include utilization and return rights) from the central to provincial/local levels and has empowered/privileged these actors to enter into networking practices with Hong Kong and/or Taiwanese investors. Their *capacities* are enhanced by the following: a) a transfer of 'utilization rights'[7] of assets (such as taxes, land, labour, loans, power supply, import/export licences, and so on) to the local level, where these rights enable local players to enter into flexible networking practices (such as subcontracting and joint ventures) with Hong Kong/Taiwanese investors; b) a transfer of 'revenue rights' (that is, the rights to possess some fruits of the utilization of assets) to the local could permit a 'second budget' as they creatively shift 'taxable' items away from the central–local budgetary accounts;[8] and c) the continual source of local fiscal softness permitted by the 'second budget' enabling local governments to 'experiment' with new local 'growth projects', external linkages, and central–local relations.

The formation of strategic networks in 'Greater China'

Capitalizing on these new capacities/opportunities, Taiwanese and Hong Kong capitalists have been drawing on their linguistic affinities and kinship ties. They have begun building socio-economic connections in the subregion. They enter strategically constructed networks with various local Chinese officials and quasi-public agencies in the subregion. These networking activities can be enhanced/thickened in and through the cultural practices of *quanxi* (relationship). When such pre-existing relationships are absent, it may take time to cultivate new linkages and this will often involve exchanging material/informational gifts,[9] taking potential partners out for dinner/karaoke entertainment, inviting them to ceremonial banquets/meetings, and making donations to the community.[10] These practices are often symbolized as gestures/signs of friendship, loyalty, mutual trust and 'giving face' (a code that communicates a sense of social importance in the network). It is through the active exchange of these gifts/favours that boundaries and significance are given to *quanxi* networks. These may consolidate a reliable and effective social space of relatives, friends, and business partners that can be called upon for utilitarian/instrumental purposes. A reliable and effective network can speed up the border-crossing time between the private and public spaces because the latter may choose to adopt a permeable boundary for the needs of the private. For example, pre-existing good *quanxi* can expedite the time needed in obtaining licences/loans/raw materials from the public/quasi-public organizations.

This account of the cultural linkages within the 'Greater China' subregion shows how a strategic group of actors/institutions is gradually emerging which is organized in a loosely hierarchical network of relations that also cuts across both the private and public and the local and cross-border domains. These networks of actors involve municipal authorities specializing in Hong Kong/Taiwan investment, county–township 'cadre entrepreneurs', as well as small- and medium-sized firms from Hong Kong and Taiwan. They draw on the pre-existing *quanxi* to build a flexible and open system in which actors can network. This allows them to build trust, obtain advice, communicate demand, and gain resources (for example, land, labour and capital) at below-market prices.

The consolidation of a subregionalized division of labour/knowledge in 'Greater China'

These strategic networks may create mutual advantages as well as enjoy support rooted in cross-border developmental communities. They represent a coalition/alliance of local party/administrative officials, their entrepreneurial affiliates, and Hong Kong and Taiwanese capital. They form the social bases of support for a *subregionalized division of labour/knowledge* for export processing in the 'Greater China' bloc. Hong Kong and Taiwan are moving up the industrial technology ladder by shifting their labour-intensive industries to low-wage and cheap-land localities in Southern China.[11] Labour is made available through de-collectivization of the countryside since 1984 and it is estimated that there are 100–150 million migrants[12] from the inland areas. In the environment of competition for capital from Hong Kong and Taiwan, coastal developmental communities compete to undercut each other by providing flexibly implemented labour laws and systems.

In response to the availability of developmental communities and 'flexible' labour regimes in the subregion, 80 per cent of Hong Kong's manufacturing industries (for example, simple electronics, toys, leather, shoe-making and watches) have moved to southern China. In the 1990s, there is a second wave of relocation which is related to the moving out of low-skill white-collar work (for example, telephone enquiry/paging service) to southern China. However, relocation of manufacturing industries remains predominant in southern China. Because Hong Kong's manufacture is largely oriented to export-processing, the transfer of the industry to southern China promotes entrepôt trade between the two regions. Taiwan's transfer of manufacture lags behind Hong Kong but the transfer of low-technology labour-intensive industries is moving forward quickly. It is reported that Taiwan's traditional industries, for instance, 80 per cent of handbag production, 90 per cent of shoemaking, and most of umbrella-making have been transferred to southern China; but a large part of raw materials, components and production equipment still come from Taiwan. Given that Taiwanese investment needs to involve a third area,[13] most of the flows of material, people, and money targeted on southern China pass through Hong Kong.

The shifting of labour-intensive processes to southern China not only consolidates Hong Kong's entrepôt role but also enhances its capacity, as a *global-gateway city*, to co-ordinate production, trade and financial services in and beyond the 'Greater China' subregion. In the case of international subcontracting and 'offshore sourcing', Hong Kong performs the function of subcontracting management. This involves sourcing, production, authority, and distribution management.[14] Thus Hong Kong specializes in producer services such as design, merchandising, production management, packaging, and logistics relevant for imports and exports. As for its role in finance, Hong Kong is an offshore financial centre for the PRC and Taiwan. The fact that the Hong Kong dollar is freely convertible and linked with the American dollar may enhance the subregion in the following ways: a) Chinese renminbi earnings can be converted into hard currency; b) Chinese flight capital can be disguised as foreign investments claiming protection and fiscal benefits; and c) Hong Kong can circumvent Taiwan's capital controls by banking trade surpluses in Hong Kong instead of Taiwanese dollars.

In the case of Taiwan, the setting up of factories in southern China has gradually transformed its operations in Taiwan into headquarters for higher-end production (for example, computer monitors, desk-top and portable personal computers, motherboards, keyboards and PC mice), R&D activities, receipt of overseas orders, procurement of materials, and provision of technical assistance and personnel training for plants in China. Economic complementarity between the partners consists of the following aspects: a) southern China possesses cheap land, trainable cheap labour, negotiable investment packages and culturally affiliated local 'cadre entrepreneurs' who are accommodating to incoming investment; b) Hong Kong is an entrepôt and global-gateway city with good global–regional connections and knowledge in finance, trade, and production management; and c) Taiwan possesses capital, applied technology and experience in administration and marketing of products.[15]

THE CO-PRESENCE OF GEOPOLITICAL DISCOURSES AND STRATEGIES

The above discussion on the geo-economic/nationalist formation needs to be linked to the co-presence of geopolitics in the subregion. In the post-Cold War era, the agenda of the US has been redefined due to Russia's increasing preoccupation with its own domestic political and economic problems. In addition, the US has had to ask other powers to help share the responsibility of maintaining world order because of its domestic decline. In this context, Japan, which is economically capable of assuming more responsibility, remains a regional partner despite its trade imbalance and economic frictions with the US.

On the other hand, the US is still inclined to act as a hegemonic power, especially in relation to emerging regional hegemons.[16] Instead of deploying the Cold-War rhetoric of 'threat', there is a shift toward an 'insurance' narrative of risk reduction in the region. In the post-Cold War era, there are certain signs and tendencies in America's post-Tiananmen foreign policy which interpret China as regional 'risks' that need to be 'insured' against and they are: a) the nationalist and assertive character of Chinese strategic culture;[17] b) the modernization of China's air and naval power;[18] c) the Sino-Russian military technology linkages; d) the Beijing–Taipei 'mini-arms' race;[19] e) China's claim of an 'historical border' in the South China Sea and missile tests across the Taiwan Strait; and f) the promotion of China's diplomatic realignments with its major Asian neighbours.[20]

For the US administration, these perceptions of China's pursuit of strategic independence and its search for an autonomous posture call for a risk-reduction strategy. This can be done by building a network of bilateral alliances with the US Cold-War partners such as Britain and Taiwan. These adopt the following strategic practices: a) supporting Chris Patten, the then Governor of Hong Kong, for his 'democratization' package, and b) selling arms to Taiwan and the symbolic act of permitting Taiwan's President Lee to visit the US. Patten's democratization project and the symbolic use of 'democratic' power attracted the attention of the Clinton administration. This was evidenced in two ways. First, in September 1992, the Clinton administration formulated the US–Hong Kong Policy. This states the principles for US policy after China's resumption of sovereignty over Hong Kong in 1997 and reasserts the importance of continuity in government programmes and intergovernmental agreements which involve Hong Kong. Second, it was prepared to take seriously Hong Kong's fears for itself about the implications of the US's China policy. This was evidenced by Clinton's reception/support of Governor Patten in April 1993 and by the repeated support for the democracy movement in Hong Kong. The emergence of this geopolitical power network is intended to communicate the collective subjectivity of Hong Kong as a democratic hub after 1997. This may become a political lever that can contribute to the control and 'otherization' of China through the signification of 'democracy' and 'human rights' symbols – a kind of power that can accentuate the growth of centrifugal (sub-state) regionalism in China.[21] This tendency for (sub-state) regionalism is also linked to the establishment of a Special Economic Zone in Guangdong. Guangdong is seen as being a relatively powerful centrifugal subregional force because of its rapid growth – its growth rate being twice the already high national average.[22] Strategically, if Hong Kong can be sustained as a democratic hub after 1997, it may become a political lever that can contribute to the growth of a centrifugal (sub-state) regionalism in China.[23] This may, in the mindset of the hegemon, induce caution in China as an emerging regional hegemon in the 'new world order'.

Parallel to this Anglo-American bilateralism is the maintenance-renewal of the US–Taiwan alliance in order to induce greater caution on the part of China. This is seen in continued US arms lease/sales to Taiwan, the initial support for the pro-independence Democratic Progressive Party (DPP), the granting of President Lee's 'unofficial' visit to the US in June 1995, and so on. The above may constitute part of the US 'insurance' strategy in inducing caution in China through the deployment of Taiwan as a lever on China's plan for 'national reunification' through the 'one country, two systems' formula.

China's national formula is not accepted by Taiwan. As Taiwanese investors began their 'family' visits to China, Taiwan's 'three nos' policy (that is, 'no contact, no negotiation, no compromise') was modified in the Guideline for National Unification in 1991. The Guidelines adopt a pragmatic attitude toward reunification and acknowledge that China is a divided nation. The KMT government holds that unification is the long-term goal which can only be achieved once the two sides of the Straits have converged towards the liberal democratic system. Prior to this, the two sides should recognize each other as equal political entities. Beijing rejects this because it believes that, once it loosens its grip on Taiwan, unification will be impossible. Under the military threat and diplomatic blockade by the PRC, Taipei is trying to normalize its foreign relations through 'pragmatic diplomacy'. This marks an effort to escape diplomatic isolation, build international contacts, and gain membership of international organizations. At the same time, it has opened the door, step by step, for people-to-people exchanges through creating the Straits Exchange Foundation (SEF) as a private, government-supported, intermediary organization for contact and negotiation with Beijing. Beijing complemented this by setting up the Association for Relations Across the Taiwan Straits (ARATS) until 1996 when China sought to influence the result of the Presidential election in Taiwan.

These strategic discourses/practices of geopolitical power communicate the strategic calculations of Cold-War hegemon in relation to 'stability' in the region. This involves deployment by the US (and Britain) of the enlightenment time-frame to narrate the security 'risks' in this region in terms of notions such as 'democracy', 'freedom', and 'human rights'. It appeals to such ideas and the values they embody in its efforts to build a network of bilateral arrangements with various allies in the region. This may take the form of 'democracy hub/movements' that can promote 'subregionalism' within 'Greater China'. At the same time, an enlightenment narrative can 'otherize' China as a 'risk' to regional stability and thereby build alliances centred upon the US. This risk-reduction strategy/mindset seeks to caution emerging regional hegemons such as China. This emerging regional hegemon is also infused with nationalist intent in remapping Taiwan within its time and space.

'GREATER CHINA' SUBREGION AS A CONTESTED SPACE

The social forces in China, Hong Kong, and Taiwan are increasingly confronted with dilemmas rooted in the co-presence of these geo-economic/nationalist and geopolitical discourses and strategies. The result is characterized by the dialectical interplay of 'multiple consciousness' which marks this subregion as contested space. These competing discourses/practices come to be (in)fused with power and ideology and thereby contribute to the *politics of identities* in the subregion. Let us start with the case of China.

Identity struggles in China

The co-presence of the geo-economic/nationalist and geopolitical discourses and strategies create moments of identity struggles in China. The coupling of geo-economics and nationalism have consolidated a strategic network with capacities for a subregionalized division of labour/knowledge. This is spearheaded by local developmental communities led by 'cadre entrepreneurs' in China. This formation has signalled the rise of (sub-state)regionalism in the south. It refers to the tendencies of provinces, counties, townships, and so on, to become more independent and signs may include the quarrels regarding the share of tax revenue to be paid to Beijing, the processes of growing independence in foreign trade, and the deviations from the policies of the central government. The rise of such (sub-state) regionalism is fuelling the identity struggles within China.

The above tendencies towards (sub-state) regionalism and coastal developmental communities intensify the struggle over China's development and nationalist identities. First, the reformer-cosmopolitan faction, which advocates greater political reform in order to facilitate the need for globalization, is in conflict with the reformer-nativist faction, which advocates the containment of the spillovers of globalization, within the central elites.[24] Such inter-factional conflict, when located at the level of subjectivity, can be seen as a struggle over China's developmental identities about what constitutes the discourse/practices/pace of 'market socialism'/'socialism with Chinese characteristics' in a globalized era. In this regard, the discursive struggles between the two factions are evident in policy debates in the Central Committee of the Chinese Communist Party. During the 1980s, the party's reformer-cosmopolitan faction (represented by Deng and Li Peng) gradually displaced 'class struggle' rhetoric with the 'coexistence of central planning and commodity production' to stimulate 'productivity', 'growth', and even 'competitiveness'. Such introduction of neo-classical discourse into central planning has met with resistance from the reformer-nativist faction (represented by Chen Yun). This group managed to reintroduce the rhetoric of 'class struggle', 'spiritual pollution', 'inflation', and even 'corruption' into the agenda.[25] This cosmopolitan–nativist struggle, when located at the level of China's developmental identity, can be seen as part of the identity politics of the subregion. Whereas the cosmopolitan faction is busily remapping China to capitalist time and the geography of global–(sub)regional development, the nativist faction is re-territorializing the development back into the socialist time frame.

Second, the emergence of (sub-state) regionalism with its pragmatic-nationalist identity may condense a north–south (or even coastal–inland) dimension to the identity struggle. (Sub-state) regionalism may resonate with the creation of a southern-oriented nationalist identity that is based on communalism and, at its extreme, separatism. This is fuelled by the increasing formal and informal contacts/linkages between socio-economic and political groups within 'Greater China' that are conducive to mutual learning and strategic renegotiation/reconstruction of identities. There is a tendency towards the condensation of a southern-oriented nationalist identity that is pertinent to the subregion/region (for example, the Hongkongization of Guangdong/Taiwanization of Fujian) and, in more extreme cases, there are claims to the revival of identity issued earlier this century for a multi-ethnic union or a federated system of government. The existence of a southern identity directly challenges the north-originated national identity largely related to the Han-centred, and centralized party–state construction. Such challenges are evident in their discursive struggles, for example, the north 'otherizes' the rise of economic power in the south as the re-emergence of 'economic warlordism' or 'feudal remnants', whereas the south (Guangdong) brands the reassertion of fiscal-investment controls by the north in 1994 as a 'conspiracy' of state-sector interests shifting central policy preference away from Guangdong to state-sector dominated provinces such as Shanghai.[26]

The emergence of (sub-state)regionalism and its southern-oriented nationalist identity is, at times, articulated with the geopolitics of the subregion. The consolidation of the geopolitical power network in the south (for example, remnants from the democracy movement in China, Hong Kong's Liberal Democratic Party, and Taiwan's Democratic Progressive Party), under the imagined leadership of the USA–UK, may orchestrate an enlightenment trope that may otherize China in and through the signification of 'risks', 'democracy' and 'human rights' symbols – a kind of power that can accentuate the growth of centrifugal (sub-state) regionalism in China. In face of such 'otherization', which may further complexify China's identity struggle, China often deploys the Singapore model of modernization as its development identity – a kind of repositioning when faced with external challenges. In this regard, China essentializes the relationship between 'modernization', 'Asian values' (for example, neo-Confucianism) and 'authoritarianism' (not democracy) as the explanation for its developmental trajectory.

The above constructions of identities are expressed in and through the narratives on 'growth–productivity–competitiveness', 'inflation–corruption–equality', 'pragmatic nationalism', 'patriotism', 'China as risk/aggressor', 'violator of human rights', 'China that can say no', and so on.... They communicated contradictory modes of collective subjectivities constructed by the different factions, leaders in the north/south, and groups in Hong Kong/Taiwan/the USA/the UK. These narratives are bound up with the (re-)inventions and emplotting of China's 'present/future' in relation to global and regional development/history. These new meanings are the reflections of power struggles in 're-presenting' and 're-making' China/'Greater China.'

Identity struggles in Hong Kong

A different geopolitical discourse and strategy privileges 'democracy'. Thus Patten as the last Governor of Hong Kong, the Hong Kong-based Democratic Party, and the US State Department have each, in their different ways, privileged Western narratives of democracy to remap Hong Kong's future. They presented 'democracy' as a universal value (especially given the recent collapse of the Soviet bloc with its authoritarian regimes) and argued that the Hong Kong people should also become the source of popular 'sovereignty' and the legitimacy of political rule.[27] Patten's 'democratization' discourse and discursive practices (for example, Governor's 'Question Time' in the Legislative Council and town hall meetings) were met with strong responses from Chinese officials. The mass-media characterized this period as one of a 'war of words'. The PRC constructed an alternative genre to counteract Patten's Cold-War rhetoric. China's reaction could be analyzed in terms of a two-part construction that clusters round 'violation' of international and constitutional documents and 'pragmatic nationalism'.[28] It revolved around two main elements, viz., historical shame and a pragmatic vision of Hong Kong's future. The construction of China's 'historical shame' derived from its defeat in the Opium Wars and the inequality involved in the Sino-British treaties which ceded Hong Kong Island and other territories from 1842 onwards. As a nationalist discourse, it privileged 'shame' related to colonialism and was used to justify unification of the Chinese nation by 'mapping' the Hong Kong Chinese as part of the 'Chinese in China' and not as subjects of a colonial power. It was through the deployment of China's 'historical shame' that China could invoke nationalism as a reaction to colonialism and used 'Chinese in China' to cultivate loyalty and patriotism in Hong Kong. On the other hand, nationalist rhetoric assumed an unorthodox form in that it was not generated, in the first instance, from within Hong Kong. Instead it was constructed by China in remapping Hong Kong into its historic territory and aimed to integrate it more fully into its economic space. In this regard, we find a contingent articulation of geo-economic/nationalist discourses such as 'open door', 'reunification', and 'historical shame' which remap Hong Kong as its 'showcase' in practising 'one country, two systems'.

The articulation of these strategies sought to redefine Hong Kong as a bastion of democracy/rule of law, as a 'window' mediating China and the global–regional networks of capital, and as a site for pursuing national reunification by attracting 'patriotic investors' to the 'motherland'. This entailed two distinctive collective subjectivities that are transmitted by competing public discourses about 'democratization' and 'pragmatic nationalism'. These discourses (re-)told and emplotted Hong Kong's 'future' in relation to different subregional memories, to enlightenment time/space, and to the geography of development. Time was recodified in the form of a temporal horizon concerned with Hong Kong's 'future' in the subregion; and space was recodified in relation to alternative trajectories concerning Hong Kong's integration into China's geo-economic and/or national time and space or its realignment with a universal and eternal value coded in terms of general democratic values, identities, and interests.

These conflicting strategies to redefine the future place of Hong Kong were the basis of uncertainties and struggles over the identities of its residents. Thus China pursued its 'second stove' strategy by setting up the Preparatory Committee and Provisional Legislature to prepare Hong Kong for its transition without having to accept the reforms proposed and implemented by Patten. It maintained this approach despite criticisms from Patten, the US and the democratic camp, and this strategy succeeded in part in undermining the last Governor's democratization project as the colony approached 1997. The significance of this strategy was highlighted when the Provisional Legislature was officially sworn in by the Chief Executive in the very early hours of 1 July to assume power from the Legco that served under Patten. Legislators from the pro-democracy camp were formally displaced from the Council and obliged to readopt their street-fighter identity battling for a democratic course.

This self-proclaimed democratic identity coexisted with an intensified version of nationalism orchestrated by the PRC. The PRC took the opportunity of the approaching 'hand-over' to remap Hong Kong within its own time and space and to homogenize memories and identify common enemies. Thus the historical time of the Chinese people and state was reunified after the parenthesis of colonial rule and the integrity of its space was reasserted through the celebration of joy and unity. This could be seen in such political-cultural events as: a) pop stars singing patriotic songs in a

214 *Subregionalism and World Order*

series of celebratory concerts in Hong Kong, Shanghai and Beijing; b) firework and laser displays to mark the return of Hong Kong to the 'motherland'; c) the participation of 'all' Hong Kong people in the largest ever karaoke festival to symbolize 'oneness' and 'unity'; and d) the showing of films and TV series on the 'Opium War' and on the resistance of people in the New Territories to colonial rule.

Identity struggles in Taiwan

In the face of its official termination of geopolitical ties with the US in 1979, the emergence of geo-economic/nationalist discourses and strategies in China in the 1980s, and the rise of an official Islander-Dominated opposition Democratic Progressive Party (DPP) in 1987, it became necessary for the mainlander-dominated KMT to reinvent new discourse/practices. In contrast to the KMT, the DPP essentializes the 'earth' identity of the Taiwanese to reinvent and recombine 'ethnic traditions' and democratic rhetoric.[29] They construct the 'One China, one Taiwan' discourse which re-presents the 'future' of Taiwan by prioritizing 'self-determination' as a means and 'independent sovereign Taiwan' as an end. In contrast to the 'mainlander–KMT construction, they isolate the identity of China from that of Taiwan by naming it as 'one China and one Taiwan'. This rhetorical construction drives a wedge between the two entities across the Strait and creates a 'self-determination' identity for Taiwan. They also map this identity to that of 'democracy' by deploying symbols/categories such as 'a new constitution' that is 'to be decided by all people of Taiwan' through 'a plebiscite'. Encoded in these symbols is the call to link the independence movement to an imagined identity derived from Cold-War rhetoric. This identity, which is derived from the articulation of the geopolitics and nationalist discourses, re-presents 'democracy' in the following manner: a) it is universal and can allow the 'marginalized' Taiwanese to (re)claim self-determination; and b) it protects them against the onslaught of communism and creates autonomy. This discursive reconstruction can be termed 'liberal ethno-nationalism' which links 'democracy' with the 'earth' identity of the Taiwanese. The creation and consolidation of the Taiwanese identity has been undermined by the dominant position of the mainlanders. Even though the KMT won strong majorities in the elections of 1986, 1989, 1992 and 1994, it has become internally divided with the growing power of 'Islander' identity demanding constitutional reform as a way of establishing an 'independent Taiwan', a direction that would depart from the

Politics of Identities Making of the 'Greater China' 215

reunification plan of China. However, the Mainlander KMT, which is mainly supported by public-sector employees, and Chinese capitalists privilege a form of 'One China' and 'pragmatic nationalist' identity – an identity that prioritizes the geo-economic/nationalist discourses and strategies but at the same time garnishes this with the security fear narrated in terms of geopolitics. Its resultant form is a hybrid of 'nationalist reunification', 'one China' encoded in pragmatism as well as a hint of 'worry about Taiwan's security'.

In Taiwan, the identity struggles are expressed in the two competing discursive formations in Taiwan and they are '(economic) reunification' (*tong*) and 'democratic independence' (*du*) of the Taiwanese at least up to 1993. They communicated two contradictory modes of collective subjectivities constructed by the DPP/US and the mainlander-KMT. These rhetorics are bound up with the (re-)narrations of nationalism, being linked with geopolitics or geo-economics. The former is structurally and discursively selective as a form of 'liberal ethno-nationalism' that articulates the deterritorialized/detemporalized nature of 'democracy' with the building of Taiwanese ethno-national identity, whereas, the latter essentializes re-territorialization and re-temporalization of China–Taiwan relations. The construction of these two imagined identities marked the moment of identity struggle which invites social and political forces to reposition themselves. In order to complexify these two discursive formations, the islander faction of the KMT, in the face of the rise of 'liberal ethno-nationalism' and the DPP, is increasingly privileging the discourse and practices of a 'Taiwanized' but, at the same time, internationalized KMT under the leadership of President Lee.

'GREATER CHINA' SUBREGION AS A SPACE FOR REIMAGINATIONS

The identity struggle not only marks the subregion as a contested space; it also marks it as a space for reimagination. Due to the coexistence of a number of reimaginations, this chapter deals with a geo-economic one that is reconstructed by the central government of the PRC. In the 1990s, the 'Greater Shanghai' (Shanghai–Pudong) discourse was constructed as a new geo-economic object for new visions in the subregion. It has been argued that this project is a counter-strategy to the possibilities of (sub-state) regionalism in Guangdong. It was represented as the object of future economic growth and is narrated as a 'tri-functional centre' (that is, centre of technology, finance, and distribution) and a 'Dragon Head for the Yangtze River'.

The building of this tri-functional centre, which aims to articulate between the global and national, involves the decentralization strategy which allowed the Shanghai Municipality to develop Pudong. Ten preferential policies for the development of Pudong were announced when it was declared as a special development area on 30 April 1990. Nine of the ten preferential policies are related to global–regional capital (for example, reducing tax rates for foreign investors to 15 per cent; and enabling foreign banks to open branches in Shanghai). These changes mark several shifts in the strategies of the central government of the PRC in the following direction: a) from the south to north as a new 'growth pole'; b) from Guangdong to Shanghai as the 'dragon head' for regional development; c) from low-tech to high-tech production; d) from old to new Shanghai (for example, Pudong); e) from small-/medium-sized firm to large-firm production; f) from familial to corporate cooperation; and g) from overseas Chinese capital to global–regional capital. In order to enhance the capacities of this new global–regional–local link, the formal regulatory pronouncements are complemented by an informal 'guarantee'. A possible set of informal institutions that induce cooperation between Shanghai's municipal authorities and central government leaders is within the central party echelon. At present, the latter is occupied by the 'Shanghai faction' (for example, the current President and Party General Secretary, Jiang Zemin, was a former Party Secretary of Shanghai) and it can reasonably be assumed that central leaders have a more realistic understanding of Shanghai's needs. These kinds of formal and informal 'guarantees' have given rise to an 'urban growth coalition' of central–local party/government officials.

The promotion in the 1990s of 'Greater Shanghai' as an emerging region with global pretensions has naturally had some effects upon actors in Hong Kong and Taiwan. They are developing complementary strategies to exploit the opportunities opening up in 'Greater Shanghai' as well as competing projects to defend their positions. Hong Kong envisions a more 'synergetic' identity. This is evident from conferences, publications and workshops organized by trade and industrial organizations in Hong Kong. At the time of writing, the discourse of 'synergy', which may project and regulate practices of association, concertation, and willingness to engage with other interests, seems dominant. They envision a new regional division of labour in terms of research, research application, production and marketing. At the research stage, Shanghai can concentrate on R&D, whereas Taiwan can specialize in applied technology and supply capital.

At the production and distribution stages, Hong Kong can supply capital, and expertise enterprise/distribution management to Shanghai. As for its financial role, Hong Kong is seen as an offshore banking centre for the PRC. It can offer loan syndication services to large state-owned enterprises to be quoted as 'H' shares in Hong Kong (for example, Shanghai Petrochemical). Hong Kong also supplied financial expertise on setting up the Shanghai Stock Exchange. In addition, the offshore location of Hong Kong also provides not only capital but helps to cycle PLA (military) capital back to Pudong to be invested in the property and stock markets. It must be noted that there are other competing visions in Hong Kong. For example, there is the remapping of Hong Kong as a 'technology centre for the Pearl River Delta/southern China'. This is evident from the recent decisions to build the Hong Kong Science Park and the 'technology corridor', which provide the symbols for the (re)imagination of Hong Kong as a 'regional technopolis' in consolidating cooperation/collaboration between overseas/local innovative firms, applied research and development in universities/research institutes in the subregion, and government-funding schemes.

In contrast, Taiwan pursues a more obviously dual approach to the new reimagination of 'Greater Shanghai' which is partly shaped by its specific structural and contingent contexts. Taiwan's increasing concern for security since 1996 and the internationalization strategy of the state seem to articulate more contradictory geo-economic/geopolitical objects. Geo-economically, it (re)imagines itself as a 'regional operation centre' or a regional hub for multinational corporations. Taiwan re-profiles itself as possessing assets such as a well-educated workforce, long-standing connections with ethnic Chinese distributors in Southeast Asia, familiarity with the languages and cultures of the US and Japan, a strategic geographical location, and an especially deep understanding of traditional Chinese culture that enables Taiwan businessmen to cope with the *quanxi* system in China. Thus, it remaps itself as a 'regional hub' that can be used by US and West European multinationals as a gateway to Southeast Asia or a much larger East Asian regional market, and especially to the mainland Chinese market. This redefinition of Taiwan's regional role is more pertinent to China's own (re)imagination of the region.

Geopolitically, security issues and the fear of over-dependence on China still rank high on Taiwan's agenda. The recent attempts by President Lee to internationalize Taiwan and the promotion of a 'southward policy' are more in conflict with China's own attempts to (re)imagine the future of the region. This can be seen from Lee's attempts to build a stronger international identity through a fresh application for entry into the United

Nations and his unofficial visit to the US in 1995; the deployment of the identity of 'Republic of China in Taiwan' has halted most semi-official talks across the Strait and even the possibility of missile attack(s) on Taiwan.[30] Likewise, the 'southward policy' aims to promote investments in Vietnam, Indonesia, Malaysia, the Philippines, Singapore and Thailand. This policy is intended to strengthen Taiwan's economic muscle while preventing its economy from having to depend entirely on China.

CONCLUSION

As this and the other contributions to this book reveal, there are several forms of subregionalism. My chapter has dealt with one such form in the making of the 'Greater China' subregion in the post-Cold War era. This exemplifies several features of that kind of subregionalism which, in East Asia, is mostly reflected in the phenomenon of 'growth triangles' or 'growth polygons' linking more or less contiguous economic spaces within a broader and more complex rearticulation of different spatial scales. For such growth triangles or polygons are being reconstituted through discourses and strategic practices on global, regional, (trans)national, and (trans-)local scales of economic, political, and sociocultural conduct in a period when it can no longer be taken for granted that one scale is predominant.[31] This is why the emergence and relative stability of growth triangles or polygons depends on the capacity of different social forces to mobilize practices and discourses to organize loosely coupled networks (and networks of networks) that cut across the private and public domains, that link immediate local and cross-border spaces to more extensive spatial horizons of action (up to the global), and that can also bridge past, present, and future.

Unsurprisingingly these emerging institutional patterns are filled with tensions and contradictions that are partly reflected in and partly caused by the politics of identities. This is why discourses and discursive practices matter. This can be seen in the ways in which geo-economic, geopolitical, and nationalist discourses and strategies and their implications for the politics of identity are combined differently in Hong Kong, Taiwan and China. But this is not to claim that subregionalism in East Asia can be reduced to a discursive phenomenon nor explained solely in terms of discourses. There are also important material bases for the emergence of cross-border regions and growth polygons rooted in actual or potential complementarities in the geo-economic and geopolitical domains and/or in the path-dependent legacies of national and ethnic histories. Indeed the strategic-relational approach advocated here should always look at actors and their strategies in relation to the differential constraints and opportunities that they face over different spatial and temporal horizons of action. For path-shaping struggles are always located within path-dependent contexts and the art of politics is to find the strategies and tactics which will enable particular goals to be realized in a pluri-spatial and multi-temporal field of action.

NOTES

1. Other changes involve retaining all above-baseline foreign exchange earned from export processing, joint-ventures and other non-trade sources. To promote infrastructure construction, the province can also exempt taxes and reduce profit remittances from enterprises in relevant sectors. The Guangdong Branch of the People's Bank enjoys greater discretion in issuing loans. The centre could still decide on tax laws and international taxation, yet the two provinces could levy and adjust all local taxes and decide on tax reductions.
2. On the idea of 'pragmatic nationalism,' see N-L. Sum, 'More than a "War of Words": Identity Politics and the Struggle for Dominance during the Recent "Political Reform" Period in Hong Kong', *Economy and Society*, February, 24, 1 (1995), pp. 73–7.
3. The Ministry of Foreign Economic Relations and Trade was formed in March 1982 as part of the reform of the State Council's organization. It merged the Commission for Import and Export Control, Ministry of Foreign Trade, Ministry of Foreign Economic Relations and the Commission for the Control of Foreign Investment merged to form this ministry.
4. For further details, see G. Shen, 'China's Investment in Hong Kong', in P.K. Choi and L.S. Ho (eds), *The Other Hong Kong Report 1993* (Hong Kong: Chinese University Press, 1993), pp. 425–54. He estimated that CITIC, a multinational under the direct supervision of the State Council, had an investment portfolio in Hong Kong worth over HKD34 billion.
5. On the power politics of local development in China, see D. Zweig, '"Developmental Communities" on China's Coast: the Impact on Trade, Investment, and Transnational Alliances', *Comparative Politics*, 27, 3 (1995), p. 268; D. Goodman, 'New Economic Elites' in R. Benewick and P. Wingrove (eds), *China in the 1990s* (Basingstoke: Macmillan, 1995), pp. 136–8; and T. Heberer, 'The Political Impact of Economic and Social Changes in China's Countryside', *China Studies*, 1, Autumn (1995), p. 59.
6. V. Nee, 'Organizational Dynamics of Market Transitions: Hybrid Forms, Property Rights, and Mixed Economy in China', *Administrative Science Quarterly*, 37 (1992), pp. 1–27.
7. For an excellent discussion of the applicability of the concept of property rights to China, see L. Putterman, 'The Role of Ownership and Property Rights in China's Economic Transition', *The China Quarterly*, 144 (1995), pp. 1047–64.

8. J. Oi, 'Fiscal Reform and the Economic Foundations of Local State Corporatism in China', *World Politics*, 45, October (1992), p. 100; and S. Wang, 'Central–Local Fiscal Politics in China' in H. Jia and Z. Lin (eds), *Changing Central–Local Relations in China: Reform and State Capacity* (Boulder: Westview, 1994), p. 99.
9. For a good discussion on *quanxi*, see M. Yang, 'The Gift Economy and State Power in China', *Comparative Study of Society and History*, 31 (1989), pp. 25–54.
10. L. Zhao and J. Aram, 'Networking and Growth of Young Technology-Intensive Ventures in China', *Journal of Business Venturing*, 10, 5 (1995), pp. 349–70.
11. In the case of sports shoes such as Nike, most of the raw materials and skills come from Taiwan through Hong Kong to Southern China. Hong Kong staff deal with designs, make sure the sample and raw materials reach the factory on time, and ship the finished products out of China through Hong Kong to their destined markets, mainly the US.
12. M. Wolf, 'A Country Divided by Growth', *Financial Times*, 20 February (1996), p. 14.
13. Before 1992, Taiwanese businessmen invested in the mainland via a third area. In November 1992, Taipei announced that if the amount involved was below NT$1 million, indirect investment in China could be made by indirect remittances and need not involve setting up subsidiaries in the third area.
14. For further details on subcontracting management, see N-L. Sum, '"Greater China" and the Global–Regional–Local Dynamics' in I. Cook and R. Li (eds), *Fragmented Asia* (Aldershot: Avebury, 1996), p. 145.
15. X. Xu, 'Taiwan's Economic Cooperation with Fujian and Guangdong: the View from China' in G. Klintworth (ed.), *Taiwan in the Asia-Pacific in 1990* (St Leonards: Allen and Unwin, 1994), pp. 142–53.
16. For further details, see J.A. Agnew and S. Corbridge, 'The New Geopolitics: The Dynamics of Geopolitical Disorder' in R.J. Johnston and P.J. Taylor (eds), *A World in Crisis: Geographical Perspective* (Oxford: Basil Blackwell, 1993), pp. 266–88; and F.M. Shelley, 'Political Geography, the New World Order and the City', *Urban Geography*, 14, 6 (1993), pp. 557–67.
17. For further details on China's assertiveness, see J. Pollock, 'The United States in East Asia: Holding the Ring' in Conference Papers on Asia's International Role in the Post-Cold War Era, Part 1, *Adelphi Paper 275*, p. 76.
18. For a fuller discussion, see J.C. Hsiung, (ed.), *Asia Pacific in World Politics* (Boulder: Lynne Rienner Publisher, 1993), p. 83.
19. For details, see C-P. Lin, 'Beijing and Taipei: Interactions in the Post-Tiananmen Period', *The China Quarterly*, 136 (1993), pp. 793–9.
20. See N-L. Sum, 'Strategies of East Asian Regionalism and Construction of NIC Identities' in A. Gamble and A. Payne (eds), *Regionalism and World Order* (Basingstoke: Macmillan, 1996), pp. 216–18.
21. See Segal, op. cit. (1993), p. 38.
22. See Segal, op. cit. (1994), p. 19.
23. See Segal, op. cit. (1993), p. 38.

24. L. Ling, 'Hegemony and the Internationalizing State: a Post-Colonial Analysis of China's Integration into Asian Corporatism', *Review of International Political Economy*, 3, 1 (1996) p. 9.
25. See Ling, op. cit., pp. 10–13.
26. T. Cannon and L-Y. Zhang, 'Inter-Region Tension and China's Reforms' in I. Cook, M. Doel, and R. Li (eds), op cit., pp. 90–94.
27. See Sum, op. cit. (1995), pp. 70–73.
28. See Sum, op. cit. (1995), pp. 73–7.
29. See T. Gold, 'Ethnicity and the Taiwanese Identity' in M. Chang (ed.), *Ethnicity in Taiwan* (Taipei: Academia Sinica, 1994), pp. 59–64.
30. See Sum, op. cit. (1995), pp. 68–70.
31. Whether this be the global (as the more fevered accounts of globalization imply), the national (as was the case in the post-war boom for Europe and North America), the local or regional (as reflected in the growing emphasis on local or regional economic development strategies as opposed to national planning), or even the urban (especially significant in the discourse of a hierarchy of world cities and the new role of cities rather than firms as 'national champions'), there are intense, continuing, and as yet unresolved struggles to re-establish the primacy of one scale over the others.

10 The East Asian Economic Caucus: a Case of Reactive Subregionalism?

Glenn Hook

The widening and deepening of regionalization processes in the context of an increasingly globalized political economy, together with the resurgence of regionalism and subregionalism as statist projects pursued by a phalanx of policy elites in the wake of the Cold War's ending, point to the growing saliency of spatial relations in the shaping of the nascent global and regional orders. In the Cold-War era, space as a source of identity was buried beneath the weight of the bilateralism at the heart of the ideological confrontation between the East and the West.[1] Now, space is being reconfigured into regions and subregions as part of the restructuring of global, regional and subregional orders. Here 'regions' and 'subregions' are regarded as the socially constituted resultants of contested socio-political processes, where state and non-state agents seek to define and redefine the boundaries amongst insiders, outsiders, and peripherals in order to produce and reproduce identities, realize interests, and acquire power in the order-building process.[2] This can be seen in the case of the proposal of December 1990 to establish the East Asian Economic Group (EAEG), which followed the Asia Pacific Economic Cooperation (APEC) initiative of just over a year earlier, thereby bringing into political and ideological contestation the boundaries and membership of 'East Asia' and 'Asia Pacific'.[3] The proposal was put forward by Prime Minister Mahathir Mohamad of Malaysia during a meeting with the Chinese Premier, Li Peng, then discussed in an Association of Southeast Asian Nations (ASEAN) working group following a meeting of the ASEAN senior officials meeting in May 1991.[4]

The outward flow of Japanese capital, technology and manufacturers by the 1980s had enmeshed the East Asian economies in an objective web of overlapping economic relationships, whether in trade, investment or divisions of labour, giving 'East Asia' a potential subjective representation as a subregional identity with political and ideological resonance, as signified by Mahathir's proposal. As the prime minister has modified his proposal in response to pressures from both inside and outside East Asia, the concept has not necessarily remained constant. For instance, the idea of including Taiwan seems to have been dropped in the face of Chinese opposition.[5] Still, the last few years have witnessed the putative membership of the East Asian Economic Caucus (EAEC) emerge gradually as 'ASEAN plus three' – China, Japan, and South Korea, with China expressing support for the idea. Even though the policy elites of Japan and South Korea have adopted attitudes ranging from being hostile at first to now being non-committal if not supportive in some statements, *de facto* meetings have taken place on a number of occasions on the ASEAN plus-three basis, even if the appellation 'EAEC' has been eschewed.[6]

Such meetings were facilitated by the change in name from EAEG to EAEC, which was proposed in the ASEAN Economic Ministers Meeting of October 1991. It was part of the process of overcoming the political opposition to the proposal outside ASEAN, as with the United States, discussed below, as well as the resistance within ASEAN, as in Indonesia's dissentient voice at the above meeting. The Group was endorsed at the ASEAN Summit the following year. By the July 1993 ASEAN Post-Ministerial Meeting the name change and consensus-building process within ASEAN had led to the emergence of an agreement within the association to accept, with varying degrees of active support, the EAEC proposal. Then, at the 1995 ASEAN Summit, Malaysia's proposal for the Mekong Basin Development and a Trans-Asia railway received ASEAN backing. Indeed, at the 1996 Foreign Ministers meeting, President Suharto of Indonesia came out openly in favour of EAEC. Moreover, by ASEAN coming to some form of *modus operandi* with APEC, meetings on an ASEAN plus-three basis could be pursued more easily by even the most reluctant of the putative members of EAEC. This was achieved at the same meeting in line with an ASEAN agreement for EAEC to become a consultative body or caucus within APEC. Finally, by agreeing to meet informally as an ASEAN plus-three group – EAEC without a political and ideological identity – the various motivations for joining in such meetings did not have to be raised as an 'EAEC' issue, as no firm commitment had been given by the participants to share the 'EAEC' identity, only to hold dialogue on economic issues with other potential member states.

Thus, over the last few years ASEAN plus-three representatives have met in a variety of fora, as at the ASEAN Post-Ministerial Conference in Bangkok in July 1994, at the ASEAN Post-Ministerial Conference in

Brunei in July 1995, and at a dinner at the ASEAN Post-Ministerial Conference in Jakarta in July 1996, when a formal agenda was established. The possibility of East Asian economic cooperation moving forward concretely also surfaced when Japan participated in the June 1996 Kuala Lumpur meeting of the East Asia Ministerial Meeting on Support and Assistance for the Greater Mekong Subregion.[7] Most significantly, EAEC, albeit still without a declared political identity, now enjoys a *de facto* existence following the first summit-level meeting between the political leaders of Europe and East Asia in Bangkok in March 1996, when ASEAN plus-three met the European powers. At the inauguration of the Asia–Europe Meeting (ASEM) top ministers of all EU members, the president of the European Community Commission, the then members of ASEAN, Japan, China and South Korea met together for the first time on an equal footing. Neither Australia, New Zealand nor the United States was involved. The second meeting of ASEM was held along similar lines in London in early 1998. In this way, although EAEC enjoys no formal existence and has been pushed proactively only by Malaysia, ASEM is emerging as an interregional economic and political forum which for the first time brings together the leaders of Europe and Asia without the presence of the United States. Indeed, ASEM provides a forum where the ASEAN plus-three (Japan, China, South Korea) formula can be put into operation, unlike in the case of APEC or the ASEAN Regional Forum, which include the United States.

MALAYSIAN MOTIVATIONS FOR EAEC

Malaysian policy elites have offered as the motivation for EAEC the ending of the Cold War and the trends towards economic bloc-building outside East Asia, as in the deepening and widening of the EC in the lead-up to establishing the European Union (EU) and the single-market in 1993, and the movement towards the signing of the NAFTA in August 1992, which crystallized fears of Mexico replacing Malaysia as the source of manufactures and as a destination for investments from some of the advanced economies. Certainly, Malaysia's immediate motivation for proposing EAEC can be attributed partly to these regional trends, insofar as the proposal emerged as a reaction to the development of regionalist projects elsewhere. Similarly, the announcement of the proposal at a time when the global-level multilateralist Uruguay Round of the General Agreement on Trade and Tariffs (GATT) seemed likely to fail as a result of the North American and European focus on economic development and trade liberalization of interest to the advanced economies, paying little heed to the needs of the developing countries, illustrates the fears of a vulnerable, export-oriented developing economy like Malaysia. As key absorbers of Malaysian exports, the advanced economies of the EU and NAFTA, especially the United States, are vital for the economic health of Malaysian exporters. An open trading system, free from protectionist measures aimed at excluding East Asian products, thus is essential for East Asian exporters. For these reasons, the EAEC proposal indeed can be characterized as a short-term defensive reaction to the threat of economic damage from regionalist projects in Europe and the Americas, on the one hand, and the failure of the Uruguay Round to address issues of central concern to the developing countries, on the other. From this perspective, the EAEC can be regarded as a way to broaden and deepen economic cooperation, trade and investment amongst a group of economies identified as 'East Asian'.

Within East Asia, however, the crucial importance of the Japanese economy to Malaysia and to the other ASEAN economies gave Mahathir's subregionalist project more the character of a metaphorical 'love call' from the 'periphery' to the 'core'. This offered Malaysia a way to try to achieve at least four goals: (1) to continue to benefit from Japanese investment, trade, technology, and so on, as one of the follower geese in the 'flying geese' pattern of economic development;[8] (2) to enhance the political bargaining power of the East Asian developing economies *vis à vis* Japan, as in dealing with any further steep rises in the price of the yen due to pressure from the other advanced economies; (3) to hinder the implementation of a divide-and-rule strategy by powers inside the region, by including within EAEC South Korea, Japan, China, and the other ASEAN members; and (4) to strengthen East Asia's hand in coping with trade conflicts and other economic issues arising between East Asia and the advanced economies of North America and Europe by pressing Japan to become more vocal as the 'voice of Asia'. Whatever these externally directed motivations, above all Mahathir could use the EAEC proposal domestically as a means to try to maintain economic growth in the face of global and regional pressures – the key to regime legitimization in Malaysia. From this perspective, the initiative can be characterized as a short-term, pro-active strategy within East Asia, but with a clear political meaning within Malaysia itself: a strategy to maintain legitimacy and power in the context of conflict within the ruling United Malays National Organization (UMNO) and potential trouble staying in power.[9]

US REACTIONS

The Malaysian attempt to create an East Asian political and ideological identity in the pursuit of East Asian interests was challenged by the United States. The American resistance to Mahathir's 'East Asian' subregionalist project can be said to have been given voice to as a way to draw attention to the connection between economic regionalization, Mahathir's project, and the ascendance of the East Asian subregion in the global political economy. For the United States is placed outside the subjective representation of an 'East Asian' identity, but at the heart of the subjective representation of an 'Asia-Pacific' identity. In this sense, the APEC can be seen as a way to exert a westward pull towards a 'Pacific' identity based on the bilateralism at the core of the Cold-War relations between the United States and Japan and South Korea, highlighting the struggle over demarcations between the EAEC and APEC projects. This issue of demarcation is epitomized by American opposition to EAEC and Secretary of State Baker's warning that the EAEG 'would disrupt the Pacific linkage that APEC seeks to build' and that Japan should not cooperate with the group (*Japan Times*, 12 November 1991; *Mainichi Shinbun*, 13 November 1991; *Nihon Keizai Shinbun*, 14 November 1991).[10] These sentiments are a symbol of American fears of Japan enhancing its regional power as well as a reflection of how the 'rapid economic growth' of 'East Asia' by now had started to erode the internal and external demarcation and image of Malaysia as first and foremost a primary-product-producing 'developing country'. Rather, as with Japan and the other East Asian economies, this non-spatial identity was being overlain with new, spatially driven economic identities, such as 'East Asian model' of development, 'high-performing Asian' economy, or 'East Asian miracle' economy.[11] Despite the recent difficulties faced as a result of the currency crises, which exposed weaknesses in the financial and economic systems of these economies and eroded the idea of them being 'miracle' economies, 'East Asia' as a way to identify these economies remains as the omnipresent metaphor.

The mad scramble made by the United States to embrace the East Asian markets as an integral part of the 'Asia-Pacific' regional identity is a manifestation of the economic interests being pursued by the US through the APEC regionalist project. In putting forward a proposal viewed by the US and other outsiders as heightening the possibility of their exclusion from the glittering prizes of the new East, Mahathir brought to centre stage the discursive struggle over demarcating boundaries within the context of the globalization and regionalization of political economy. Thenceforth 'model of development', 'culture' ('race') and 'economic bloc' have been privileged in the discourse on the East Asian project. Indeed, in proposing boundaries to EAEC based on an 'Asian' identity, which led to the exclusion of Caucasians ('white') Pacific Asia,[12] Mahathir was able to reverse the pre-war 'inclusion–exclusion' debate of the League of Nations, when Japan was excluded from equal treatment politically due to 'colour', with his own project now being tarnished with the brush of 'bloc' and 'culture'. In this sense, the EAEC proposal is quintessentially political and ideological in nature.

EARLY-STARTERS AND SPRINTERS

The West's racist response to the Japanese emergence as a power to be reckoned with at the global level offers an insight into how the early-starters dealt with the political ascendance of a late-comer on the road to modernization and economic development. In the global political economy of the 1980s and the 1990s, the relative economic decline of the United States and the challenges posed to the advanced economies by the international competitiveness of the East Asian sprinter economies has cast into sharp relief the question of the present-day political defence against a similar sort of challenge. If not only Japan, but other East Asian economies, are set to eat into sector after sector of the US and other advanced economies due to the international competitiveness of their exports, discursive as well as extra-discursive strategies will need to be deployed in response. In traversing a different catch-up road of development, the East Asian economies have not only been attacked for being 'different', but some of their policy elites now seek to defend this difference on their own terms. In the context of the emerging orders of the post-Cold War world, which provide opportunities as well as set constraints on the pursuit of the EAEC subregionalist project, Malaysia has gone beyond the post-1945 Japanese strategy of an economic, political, ideological, and security orientation towards the Pacific, as part of Japan's Pacific seaboard development centring on the United States, and has sought rather to legitimize 'East Asian culture' and the 'East Asian model', and deny the charge of creating an 'economic bloc'. This has not gone unchallenged.

Discursively, the US has responded to the East Asian economies by branding their international competitiveness as inherently 'unfair'. It is only possible, the argument runs, due to 'social dumping': the low wage levels of East Asian workers, their lack of rights, the state's infringement of human rights, the government's lax regulation of environmental polluters, and other negative manifestations of the 'culture' and 'model'.[13] American threats to restrict East Asia's 'unfair' imports and to remove the 'advantages' of the Generalized System of Preferences (GSP) seek to reinscribe the discourse of development. As in the case of earlier Western criticisms of Chinese coolies willing to work for 'unfairly cheap' wages in the Meiji era, when Japan started to pursue its catch-up policy, East Asian sweat labour and an environmental free-for-all are seen to offer 'unfair advantages' in a globalized political economy, where quick-footed multinationals relocate in order to take advantage of these conditions and the GSPs.

Similarly, the 'East Asian model' has been attacked for being over-reliant on the 'unfair advantages' accruing through the intervention of the state in the economy, rather than left to the 'fair', yet invisible, hand of the market; for pushing export-oriented growth in order to take 'unfair advantage' of the open markets of the advanced economies, rather than nurturing the domestic market; and for the domestic market being closed to 'fair' international competition, rather than open to all-comers. What is more, these trade-or-die strategies pursued by the new East Asian sprinters are characterized as being possible only due to the preparedness of the US government to sacrifice the youth of America in order to protect the international peace – the metaphor of the American 'policeman'. By benefiting from the international peace achieved by the US these economies are said to have been able to take a 'free ride' at America's expense. From the 'Japan problem',[14] the discourse has widened to the 'East Asia problem'. In other words, these late-starter, East Asian sprinter economies are attacked for gaining unfair advantages due to being 'different' in terms of both 'culture' and the 'model' of development being pursued, on the one hand, and for taking 'unfair advantage' of the presumed 'public good' of US-provided security, on the other.

The validity or not of such criticisms need not detain us here. Our interest is rather in the Malaysian discursive response. For in the discourse on economic development and the EAEC subregionalist project, the 'East Asian' model, 'Asian culture', 'Asian human rights', and so on, have been defended in all their 'differences'. In other words, the western attacks have been viewed not simply as a wailing cry of 'unfair' by early-starters seeking to slow the pace of the East Asian sprinters, but rather as a challenge to their very right to define their own existence – the political right to be 'different'. For Mahathir, 'European nations and the US have a reputation of using a host of issues like human rights, trade unionism, exchange rates, media treatment, environment protection and democratic practices to suppress the economic growth of potential competitors.'[15] This quite clearly signals a shift in discursive strategies, away from 'universalism' (western) to 'particularism' (Asian). In this sense, globalization and universalism, on the one hand, and regionalization and particularism, on the other, now are intersecting in the discourse on East Asian development and the EAEC subregionalist project.

What this means is that, in proposing EAEC, Mahathir was seeking to make the group the vector for a shared political and ideological identity embracing a particular geographic space regularly articulated by insiders, outsiders, and peripherals as being inherently different. Indeed, the emphasis on difference *within* this space – in terms of civilizations, cultures, religions, levels of development, political systems, and so on – has been in the past the hallmark of the discourse of the putative group making up EAEC. Even though the demarcation of the boundaries to this subregionalist project has been tentative at best, the infusion of an East Asian model of development and particularistic Asian culture, not just economic indicators, into the interstices of this new identity, point to the highly political role the proposer of this economic caucus is seeking to play. In a similar vein, the use of particularism to this time exclude Australia and New Zealand from a newly proposed forum for dialogue on East Asian security as can be seen in the position jointly taken by President Ramos of the Philippines and Mahathir that these two countries 'are not used to the Asian way of dialogue. They adopt a position with major differences to Asian countries in regard to democracy and human rights.'[16] In other words, despite a nationalist Australian politician's fear of the nation being 'swallowed by Asians' (*Asahi Shinbun*, 18 October 1996), both New Zealand and Australia are denied inclusion as 'Asian countries', excluding them from 'East Asian' dialogue on security and economy, as neither are seen to share particularistic 'Asian' values.

In terms of extra-discursive strategies, two of the most potent in the US arsenal are to implement policies in order to change the balance of economic advantage between the United States and East Asia, on the one hand, and to set up obstacles to the realization of the EAEC subregionalist project, on the other. The former case is exemplified by the gradual removal of the GSP: in 1985 duty was imposed on micro-wave ovens imported from Singapore; in 1987 the GSP was removed from 250 items imported from Taiwan, South Korea, Hong Kong and Singapore; and in

1989 the Newly Industrializing Economies (NIES) were excluded totally from the GSP, leading to fears in Malaysia that it also might be targeted as a way to reduce the economy's competitiveness in the US market.[17] In the latter case, as Japanese cooperation is crucial to the successful realization of EAEC, the United States apparently has sought, through a secret agreement, to bind firmly the Japanese government to a political commitment not to join EAEC (*Mainichi Shinbun*, 29 November 1991). Given the ideological power of bilateralism to continue to influence the Japanese policy elite, US opposition has served to impede Japanese political support for EAEC *qua* EAEC. Even though ASEAN-plus three meetings have taken place, therefore, the United States has been able to call on the Cold-War legacy of the 'security-treaty card', as in the redefinition of the US–Japan Security Treaty in 1996 and the proposal for new Guidelines for Defence Cooperation in 1997, [18] as a way to modify Japanese behaviour in respect of trade and the economy. American actions have slowed down, if not fully stopped, the political and ideological integration of Japan into the subregionalist project *qua* project. At the same time, the United States increasingly has sought to act as the prime mover in APEC, as seen at the 1993 Seattle meeting, in order to institutionalize an 'Asia Pacific' regionalist project linking the US economy to the economies of East Asia. It seeks in this way to marginalize the EAEC project and prevent the possibility of an East Asian economic 'bloc' emerging.

For Malaysia, extra-discursive strategies naturally have been rooted in attempts to gain the support of the key members of the subregionalist project, ASEAN, Japan, South Korea, and China. In making the EAEC announcement during a meeting with Premier Li Peng, for instance, Mahathir sought to ensure that China would support the proposal. In the case of Japan and South Korea, Mahathir and other Malaysian ministers made visits to Tokyo and Seoul several times in the early 1990s, albeit without apparently being able to convince the Japanese to offer open support, although South Korea had offered tentative support by 1993. In the case of ASEAN, the Malaysians were able to gain the association's backing for EAEC through political negotiation and compromise. The centripetal forces of integration within ASEAN were facilitated by the association's increasing resistance to the US's pursuit of its own interests through domination and the increasing institutionalization of APEC. In this way, Malaysia has taken concrete actions in order to try to win the backing of these countries, although the emergence of the ASEAN-plus three formula points to the difficulty of a weaker country successfully realizing a subregionalist project opposed by the United States.

MALAYSIA AS A LATE-COMER

The difficulty faced by Malaysia in promoting a subregionalist project is linked to the deeper structural problem of a late-comer. More specifically, this subregionalist project needs to be understood within the context of the East Asian economies as late-comers in the global scramble for economic development and political survival. This draws our attention to the dynamic interrelationship between Japan as an 'early' late-starter, the Newly Industrializing Economies (NIES) and the ASEAN sprinters as 'middle' late-starters, and China as a 'late' late-starter. In its own attempt to catch up with the Western early starters, the Japanese advanced into East Asia, thereby giving concrete expression to a regionalist project rooted in imperialism – an option clearly not open for the next-comers. The empire sought to subordinate, both physically and ideologically, other Asians within an exploitative, hierarchical structure – the Greater East Asia Coprosperity Sphere. While the failure of this imperialist project meant that the 'middle' and 'late' late-starter victims were able to free themselves from the shackles of both western and Japanese imperialism, the legacy of especially neighbouring Japanese imperialism left the later-comers with the need to address a number of important issues in their own post-war struggle for economic development. Malaysia was in a position to address these issues as related to Japan.

First is the legacy of Japan's aggressive war in East and Southeast Asia. It has emerged in a number of forms over the past half-century: in the demand for war-time reparations, which coloured relations between Japan and the empire's victims during especially the 1950s; in the clarion call for a clear-cut apology for war-time aggression, which resounds periodically in response to conservative politicians who come out in praise of the benefits of Japanese-style imperialism; in the criticism of ministerial visits to Yasukuni Shrine,[19] which were raised in the wake of Prime Minister Nakasone Yasuhiro's decision in the 1980s and other political leaders' decisions in the 1990s to pay homage to the interred spirits of Japanese war criminals; in the attack on the whitewashing of history in school textbooks in the 1980s,[20] which obfuscate the concrete details of the empire's aggression; and in the claims for compensation for the Asian and other 'military sexual slaves' used by His imperial forces, which has led to cases of these women seeking compensation from the Japanese state in the 1990s.[21] This imperialist legacy, together with the recurring problem of leading Liberal-Democratic Party (LDP) politicians trotting out ahistorical justifications for the empire's war-time deeds, has been a thorn in the side of those Japanese policy elites seeking to play a fuller role in the region, beyond narrowly economic affairs.

Such a role was made easier in 1994 when, on a visit to Malaysia by Prime Minister Murayama, Mahathir stated: 'We prefer to look towards the future rather than harp on actions in the past. I cannot understand why the Japanese government keeps apologizing for things that happened 50 years ago.'[22] In proposing a subregionalist project including Japan, Mahathir is not looking to the history of imperial aggression, but to a rose-tinted economic future where Japan would serve as the economic hub in the EAEC wheel of fortune. By seeking to lay to rest the memory of the imperialist power whose war-time soldiers plundered the region and whose post-war conservative leaders periodically legitimized their actions, he was seeking to address an especially prickly issue – the question of war responsibility. It was easier for Mahathir to call on the Japanese government to 'stop apologizing', as the Malays, on balance, suffered far less at the hands of Japanese imperialists than say the Chinese or Koreans, who continue to appeal to the legacy of the war in their dealings with Japan. Similarly, political leaders in Malaysia have not used the war experience as a political tool domestically, as in the case of both China and South Korea. At the same time that Malaysia thus found it easier to make a 'love call' to Japan, Mahathir's subregionalist initiative was in keeping with the new definition of Japan's own, more pro-active role in the region, which required a forthright treatment of war responsibility in order to be pushed forward. The apologies for wartime actions made by prime ministers Hosokawa and Murayama following the collapse of LDP rule and the emergence of coalition governments in the early 1990s should be seen in this light. In this way, for Japan to play a central role in EAEC, the legacy of the war needed to be addressed. Mahathir's proposal was made at this time of change in Japan's own attempt to redefine its role in the region.

Second, the fear of a recrudescence of the East Asia Coprosperity Sphere at the heart of Japanese imperial expansion, if in the neo-imperialist rather than classic-imperialist style, has haunted the empire's victims for much of the post-war period. In seeking to integrate Japan into the post-war capitalist order, US policy-makers designated Southeast Asia as the source for Japanese raw materials and as a market for Japanese manufactured products – the 'periphery' for Japan as the US's 'semi-periphery'.[23] This served to enmesh these economies in a subordinate economic relationship with Japan in the early years of post-war development, as did the reparations made from the mid-1950s onwards. The same can be said for Japan's Official Development Assistance (ODA), first offered as a political response to 'East–West' concerns, as at the time of the 1965 normalization of relations with South Korea, and then later as a response to 'North–South' concerns, as after the 1973 oil crisis. Overwhelmingly, Japanese ODA has been directed towards East and Southeast Asia. The waves of investments that started to wash across the region, especially following the breakdown of the Bretton Woods system, brought Japan's further economic domination of the region in the 1970s. In the face of such an economic onslaught, it was not long before Asian nationalism flared into the open. In the violent student protests in Thailand and Indonesia at the time of Prime Minister Tanaka's 1974 visit, the widespread concern over the 'neo-colonial' inroads made by Japanese corporations was fused symbolically with the search for national identity, which was seen by especially the intellectuals and students of Southeast Asia to be threatened by the influx of Japanese goods, money and ideas. In other words, as Japan moved deeper and deeper into the economic life of East and Southeast Asia, the Japanese 'economic animal' clashed with the forces of indigenous nationalism.

Nevertheless, the anti-Japanese demonstrations and the calls for a boycott of Japanese goods by Thais and Indonesians were in marked contrast to the response in Malaysia, where links with Japan did not suffer to the same extent the brand name 'neo-colonial', nor was the Japanese presence such a threat to national identity, which was emerging in the context of Malaysia's multi-ethnic society, 'Bumiputra' policy – that is, a policy favouring Malays or *bumiputras* (indigenes), and continuing close ties with Britain. For such reasons, Malaysians harboured less fear of economic domination by Japan than in some other parts of East and Southeast Asia. It was thus a Thai minister, rather than a Malaysian, who lamented in 1973 that 'the Japanese are coming ... by Japan Air Lines, using Japanese made taxis, staying in a hotel built by Japanese capital and under Japanese management, eating in Japanese restaurants and employing guides of Japanese origin'.[24] It thus was much easier symbolically for a political leader in Malaysia, rather than in Thailand or Indonesia, to propose a key political role for Japan in a new subregionalist project. What is more, it was much easier for Thailand, Indonesia and other members of ASEAN to accept the proposal after economic growth and a boost in confidence in the years since the Tanaka protests had created a new middle class seemingly more interested in emulating the life-style of an oniomaniacal Japanese consumer than in boycotting, banning or burning made-in-Japans as symbols of 'neo-colonialism'. In other words, Malaysia was able to deal with another issue better than others in the region, especially as the former colonial master, Britain, was still seen by many to be central to the nation's economic well-being.

Third, if the Japanese can do it, why can't we? Despite these cacophonic memories of war-time classic imperialism and protests against post-war neo-imperialism, by the 1970s the spectacular economic growth of Japan had bedazzled the nationalist, authoritarian leaders of the newly developing nations of East and Southeast Asia, despite the ideological pull to the American heartland. They shared with their congeners in Japan an anti-communist, pro-capitalist ideology rooted in a belief in the strong state. Some were willing to be tempted more than others by the East Asian 'economic model' offered by Japan. In pursuing a 'Look East' policy from 1982,[25] Mahathir declared openly his goal of emulating the Japanese success, suggesting that, at least insofar as economic models are concerned, Malaysia had no 'Japan problem'. In this sense, the 'Look East' policy was a precursor of EAEC in seeking to carve out for Malaysia a 'special relationship' with Japan.

Thus, in setting out to 'learn from Japan', Malaysia was turning away from a western model of development, choosing from the vast array of Japanese values, norms, and practices a limited number as quintessential to economic success, such as the work ethic, social discipline, in-house unions, management techniques, and so on. Whatever the reasons for Japanese economic success – the roots of modernization in the Meiji period, the beneficial international environment, the economic bonus redounding from the Korean and Vietnam wars, the low spending on the military, as well as the sorts of domestic practices and values praised by Mahathir – the aim was to follow a different path of development from the one laid down in the West. At the same time, however, Malaysia's attraction to the Japanese model occurred at precisely the time that the Japanese themselves were questioning the costs of the model of development that they had adopted, as seen most cruelly in the victims of industrial pollution and environmental disasters on the domestic front, and the trade conflicts with the advanced economies on the international front. The response from Japan was thus less than Mahathir had hoped for. By the end of the 1980s, after the Malaysian economy had been buffeted by the impact of the sharp rise in the yen following the 1985 Plaza Accord currency realignments amongst the advanced economies, the 'Look East policy' gave way to EAEC. In this Mahathir sought to maintain a central role for the Japanese economy. The proposed subregionalist project was in essence a way for a 'middle' late-starter to try to 'catch up' by emulating the model of an 'early' late starter within the emerging global, not simply national and regional, political economy. In borrowing selectively from the Japanese economic model the ruling party in Malaysia was seeking to legitimize the regime's economic policies in an 'East Asian' context; by proposing EAEC, Mahathir was placing these policies squarely in the context of globalization and regionalization, continuing to rely on Japan in pursuing economic development. For Mahathir, at least, the benefits of continued dependence on the Japanese economy apparently outweigh the costs.

Fourth, given the shared interests and needs of late-starter, export-oriented East Asian economies, could Japan not act as the 'voice of Asia,' despite the legacy of the war and the residual fear of Japanese economic domination? Mahathir obviously thought so. International economic fora like the G-5, G-7 or G-10 are in the hands of the advanced economies of North America and Europe, with Japan the sole voice of Asia. In speaking out in favour of open markets in North America and Europe, Japanese political leaders are taking a stand against the protectionist forces of harm to Malaysia as well as to Japan and the other export-oriented East Asian economies. Similarly, when Japanese policy elites take a stand against American pressure to open more widely the Japanese market,[26] they are resisting the same sorts of pressures feared by the elites in other East Asian economies, which cannot bear the international winds of competition in certain, if not all, areas of economic life. Nevertheless, Japanese policy elites still are locked overwhelmingly into the bilateral ideological prism of the Cold War. Thus, in coauthoring *'No' to Ieru Ajia* (The Asia That Can Say 'No') with Ishihara Shintaro,[27] the nationalist politician famous for writing *The Japan That Can Say 'No'*,[28] Mahathir was seeking to raise his own voice in the Japanese 'Asian' discourse. As a small, developing economy which relies on exports, Malaysia can most effectively raise the decibels internationally as part of a group, rather than as a solivant voice. Without Japan, the East Asian voice would be like crying in an international wilderness: building up political support amongst the new Asian nationalists in Japanese policy-making circles is thus a way to strengthen the Malaysian hand and help promote Japan's reasianization.

In this sense, the new trends towards regionalism and subregionalism in the 1990s offered a way for Malaysia and the other East Asian export-oriented economies, which share many common characteristics and problems with Japan, to bolster their international bargaining power by working as a subregional group, rather than dealing on a one-on-one basis with the big powers. If Japan could be persuaded to play a fuller role in international fora as the 'voice' of this East Asian group, then Malaysia would be in a stronger position, too. In this respect, the EAEC can be regarded as a classic strategy of the weak seeking to join together with one of the strong which shares some of the weak's interests and needs, in order to respond more effectively to the pressures arising from regionalization and globalization in a global political economy still dominated by North America and Europe.

The final issue relates to China rather than Japan: the role of the overseas Chinese in Southeast Asia. Even more than for 'middle' late starters, 'late' late starters face the difficult task of capital accumulation in the pursuit of rapid economic development. One option for China is to pull in the overseas Chinese to invest in the 'motherland' by exploiting the shared ethnic identity of the diaspora, especially those of the ethnic Chinese in a Southeast Asian country where the Bumiputra policy might have restricted their economic opportunities. Similarly, with the Chinese economy enjoying rapid economic growth, suggesting the possibility that the giant socialist market economy also might become a showcase of the 'East Asian miracle', the shared sense of identity between many diasporic Chinese and those in China exerts a strong push towards the economic opportunities in the now-booming mainland. What this means is that, for ethnic Chinese in Malaysia, both 'pull' and 'push' factors were in the late 1980s and early 1990s starting to direct investments towards China, rather than towards the economy at home. It is true that some of these investment outflows actually found their way back into the home economy, as in order to circumvent the restrictions imposed on the Chinese role in the economy by the New Economic Policy, Malaysian Chinese sent funds to Taiwan and Hong Kong and then transferred them back into the country as foreign direct investment. Nevertheless, with overseas investors also becoming attracted to the Chinese rather than the Malaysian economy, the possibility of capital starvation could not be totally ignored by the Malaysian policy elite.

As in the case of other Southeast Asian economies, the ethnic-Chinese continue to play a pivotal role in Malaysia's economic life, despite the attempt by the ruling party to expand the part played in the economy by the Malays through the New Economic Policy. If the net capital outflows from the ethnic-Chinese could not be stopped, the dilemma for the Malaysian leadership is plain to see. Although only around 30 per cent of the population is ethnic-Chinese, their continued domination of the Malaysian economy means that any capital diverted from investment in Malaysia to investment in China carries disproportionate meaning. In this sense, the announcement of the EAEG at the time of Premier Li Peng's visit indicated that, although Japan was expected to play a key role as the voice of Asia in the Western-dominated fora of the global political economy, fears of the negative effect of such outflows also was of concern – hence the need to gain Chinese support for an East Asian regionalist project embracing both China and Japan. In the face of the economic pull of the Chinese economy, the Malaysian leadership thus sought to play a coordinating role in competing for capital in order to continue economic development at home.

EAEC AND POST-COLD WAR SECURITY

In East Asia as well as elsewhere, the ending of the Cold War and the pressures of globalization and regionalization have set the opportunities and constraints in pushing forward with a subregionalist project like EAEC, whatever the motivations and issues involved. At the same time, however, as the start of the end of the regional Cold War in East Asia can be traced back to the normalization of Sino-American relations and the ending of the Vietnam War in the 1970s, the peace settlement in Cambodia, together with the successful economic development of East Asia, can be said more specifically to have set the regional opportunities and constraints for this subregionalist project. In other words, the above focus on the economic, political and ideological implications of EAEC should not divert our eyes away from the security implications of economic subregionalism in the post-Cold War world. This is not only in the sense of the importance of a stable security environment for moving forward with economic cooperation. It is more broadly in the sense of the widespread recognition of the declining utility of military force in effectively and economically resolving security issues in the emerging regional and global orders, and a widespread, though not total, rejection of the apotheosis of military solutions to human problems. It is even more rooted in an understanding of security as embracing social, economic, and environmental, not just military, dimensions. This means that EAEC and the ASEAN-plus three meetings on economic issues can be viewed as part of the emergent yet still inchoate means for dealing with certain security issues on a multifunctional, multilevel, and multilateral basis.

In this way, the political space for an EAEC-type initiative can be said to have emerged in the triple context of successful economic growth, the end of the Cold War, and the peace settlement in Cambodia. For ASEAN, the Vietnamese withdrawal from Cambodia in 1989 and the United Nations agreement on the Cambodian peace settlement in 1990 offered the opportunity to include Vietnam in the association. This was one way by which ASEAN could give renewed meaning to its existence: by expanding the association's boundaries outward, beyond the East–West divide of the Cold War era, Indochina could be brought gradually within ASEAN's embrace. At the same time, however, the Japanese role in the economic reconstruction of Cambodia and the economic potential of Vietnam could be expected to redirect ODA and investment away from some of the older members of the association, thereby linking economic and security issues. The EAEC proposal served to draw attention to one of these older members, Malaysia, now no longer a major target for Japanese aid, but still in need of Japanese investments.

Second, even if EAEC does not function explicitly as a forum for dialogue on security issues, as in the case of the ASEAN Regional Forum,[29] meetings between ASEAN, along with China, Japan, and South Korea, serve a confidence-building function. It is true that, as far as institutionalization is concerned, the ASEAN Regional Forum is emerging as the multilateral forum for discussing security in East Asia and Asia Pacific more widely. Nevertheless, as the putative EAEC is made up of East Asian nations, building confidence within the same group through EAEC-like meetings serves to turn gradually EAEC into the economic launch platform for East Asian security dialogue. There are some issues, such as territorial disputes between China and members of ASEAN, which might be dealt with best outside a framework involving the United States. Indeed, in order to move to settlement of territorial problems, multilateral dialogue involving China is essential. This is precisely one of the purposes of the proposal put forward by Ramos and Mahathir mentioned earlier: a security version of EAEC. Like EAEC, it similarly seeks to be exclusionary by appealing to a discourse of 'Asianness', as Ramos suggests: 'by using modest and thoughtful language and through harmony amongst us, Asians have built stability in the region'.[30] From this perspective, Asian values appear as the source of stability in East Asia and the source for the resolution of 'Asian' security issues.

Third, with the ending of the Cold War, and with closer relations now in place between China and Russia, the regional big powers are redefining their roles in the region, especially as the earlier Soviet pull-out from Vietnam and the break-up of the Soviet Union basically spelled the end of Russia as a key military actor in East Asia. How can the small, weaker powers like Malaysia respond to the end of the era of heavy regional involvement by the big powers? With the reduction of US forces in East Asia and the 1991 vote in the Philippine senate to terminate the lease on Subic Bay and Clark air base, ending the US presence in 1992, a policy of balancing off Japan and China with the military weight of the United States was bound to face difficulty. Such a balancing off was made even more difficult by the East Asian resistance to US economic influence, and domestic pressures in the US to cut back on the military. The Chinese proclivity to revert to hard power tactics in East Asia, as in the use of force in disputes over territory, meant little support could be expected from the smaller East Asian powers for the present Chinese naval build-up, even if this could be seen as a way to fill any perceived vacuum left by the United States. The Japanese economic presence has been viewed as an essential balance to China, but fears about the Japanese taking on a full military role remain.[31] So the Cold-War formula of American military power to contain China, Chinese political power to balance Japan, and Japanese economic power to balance the two, no longer seems viable in 1990s East Asia. In the newly emerging East Asian order, seeking to bring together key players within the same grouping, rather than seeking to balance one off against the other, thus could serve to reduce tensions arising from the reduction in US forces and the Chinese military build-up.

This brings us to the fourth point: the 'China problem'. Clearly, in realist terms, Malaysian and wider ASEAN concerns about the power and ambitions of China serve to draw the smaller Southeast Asian nations together. In this sense, ASEAN fits the classic realist evaluation of the motivation for the development of regional cooperation, where material capabilities and size lead nations to cooperate in order to balance a bigger power. But as Mahathir states:

> In my view, to build our East Asian Peace on the basis of a balance of power is not possible. It is not advisable. And it is not productive of the warm, co-operative and enduring peace we must work for.

> The reason it is not possible is because most of us cannot afford the enormous expense that would be involved. Can we all build military machines that can balance the military capabilities of China?

> ...I can hear the response of the undeterred Balance of Power enthusiasts: if no single nation can create a Balance of Power on its own, create alliances. But who will agree to create alliances against China?...against the United States?...against Japan?...against Indonesia?...[32].

From this perspective, the EAEC subregionalist project can be seen as a means to embed China in an overlapping network of political and economic relations, which in and of themselves do not guarantee, but can possibly contribute to, the creation of a more stable security environment in East Asia – an alternative to the balance of power. Even as the ASEAN plus-three, the EAEC functions as one layer in the complex and overlapping layers of dialogue, cooperation and networks which, with the globalization of political economy and the end of the Cold War, are emerging in different regions and subregions of the globe.

Finally, in the cases of EAEC, the ASEAN Regional Forum, and even APEC,[33] Malaysia and other members of ASEAN are emerging as the focus of the East Asian network of overlapping connectednesses, suggesting the transition of ASEAN from being inward-looking to being more outward-looking. The association's role in each of these groupings suggests that, in the search for new global, regional, and subregional orders, agents at different levels of 'globalness' and 'regionalness', and with different agendas, are emerging to complement, though not replace, the national, international and supra-national levels of governance now in place. Whether or not EAEC can be regarded as an emergent illustration of one of these new levels of governance will be returned to below.

CONCLUSION

The proposal to establish the EAEC can be characterized in a narrowly economic sense, as a reactive strategy seeking to respond to the triadization of the global political economy as well as a proactive strategy seeking to establish Malaysia's position within an East Asian political economy dominated by Japan. It can be seen as a subregionalist project by Malaysia to coordinate roles in an East Asian division of labour based on the existing and emerging comparative advantage of the putative members, tying the life-line of the Malaysian economy more closely to the dominant regional economic power, Japan. Working in tandem with Japan would enable Malaysia to play a crucial role in moderating the economic tensions arising from the Japanese economic presence and to be in a better position to consult with Japanese policy-makers in the face of another devastating rise in the value of the yen. As one of the geese in the metaphorical 'flying geese' pattern of development, Japan would continue to play the key role in the Malaysian and other East Asian economies by using trade, foreign direct investment, technology, and ODA as a way to maintain a mutually rewarding, if not always equal and harmonious, East Asian political economy, structuring the 'cooperation–competition' complex to subregional advantage. The uneven development and the 'North–South' disparities in East Asia thus would be tackled within a subregionalist framework, seeking to build East Asian strength in competition with the other two triadic cores of the global political economy, North America and Europe. As the leader of the group, the Japanese would be expected to coordinate the division of labour, seek to eliminate over-competition amongst follower geese, and play a role outside the subregion as the 'Asian voice' seeking to resolve trade friction between East Asia and the other two triadic cores. In this context, the Malaysian proposal can be understood within the global context of a move away from the national scale of organizing economic life to the regional and subregional, where East Asia would be strengthened in order to deal with extra-regional competitors, especially North America and Europe.

The approach taken here has been to examine the emergence of this subregionalist project within the broader structural context of how the early-starters and late-comers have responded to the changes in political economy signalled by globalization, regionalization, and the end of the Cold War, paying attention to the discursive practices and discourses at the heart of the struggle over demarcating boundaries of inclusion and exclusion. It has treated spatial relations and the imputation of space with identity as a contested socio-political process involving competing agents and interests. This is clear in the comparison between the APEC and EAEC projects, with the former seeking to puncture any perceived spatial congruence between 'culture' and 'region' in the latter through the promotion of an 'Asia Pacific' identity. The discursive practices at the heart of constructing an East Asian identity highlight how Asian culture and values have moved to centre stage in the discourses on this subregionalist project. But a simplistic 'clash of civilizations' or 'clash of cultures' is not regarded as symptomatic of the present era of transition,[34] as 'culture' and 'values' are viewed as contested within as well as amongst different subregions and regions, cultures and civilizations. It is more to suggest that, with the end of the Cold War and the decline, if not elimination, of the military response to the economic ascendance of late-starters, culture and values are being used by policy elites and others in order to impute space with subregional and regional boundaries and meaning as part of a strategy to cope with globalization and regionalization. Given that, in 1990, EAEC was greeted with open hostility by the United States, the ASEAN plus-three meeting with Europe in ASEM indicates the gradual, though hardly total, success of the Malaysian project. At the minimum the strategy has contributed to the emergence of linkages between the East Asian and the European wings of the triad, excluding the North American. In other words, for the first time East Asia has been able to play the 'European card'. Indeed, the early-starter, former colonial masters of Europe in 1996 sat down with the 'coloured' late-starters of Asia on the basis of equality for the first time in history. In this way, the early-starters of North America and Europe have been forced to respond to a subregionalist project proposed by an East Asian late-starter, even if one of the immediate motivations for the proposal was as a reaction to the strengthening of regionalism in the advanced twin centres of the triadizing global political economy.

Still the above incantation of the economic, political, ideological and security implications of East Asian subregionalism in the 1990s should not inure us to the longer-term trends in building regional and global orders. For although the EAEC can be considered to be reactive subregionalism at one level, at another level the action–reaction process set in motion by the regionalist projects of North America and Europe is dialectic, with roots planted deeper in the processes of economic development and transformation in the global political economy. At various times, the challenge to the established hierarchy of the world order posed by the late-starters has been met with 'hard' rather than 'soft' power. With the end of the Cold War, however, carrots are easier to legitimize than sticks. Is, then, a transition from a hegemonic mode of global and regional orders now underway, with new levels and sources of governance emerging in complex, overlapping and inchoate ways, with EAEC a harbinger of things to come?

Looking back from the twenty-first century, EAEC may appear as nothing more than a speck, a modest subregionalist project proposed by a late-comer on the road to being integrated into the new hegemonic order. Alternatively, it may have disappeared, being symptomatic of the breakdown of the regional and global orders into widespread, unremitting anarchy. Or, it may live on and be transformed into one of the new levels or layers of overlapping governance, where new mechanisms have been established on multiple levels (supranational, international, regional, subregional, national, and so on), in multiple fields of action (politics, economics, culture, security, and so on), within multiple sites of authority which, as a result of the de-statization of authority in the process of responding to the opportunities and constraints of the emerging global, regional and subregional orders, are now taking root gradually in complex combinations amongst these levels and fields of action.[35] None of these is pre-ordained for the twenty-first century. Nevertheless, the strategies and actions of policy elites pushing EAEC, even if viewed as nothing more than symptoms of reactive subregionalism in the short term, might in the longer term contribute to the shape of things to come.

NOTES

1. On the case of Japan, see G. Hook, 'Bilateralism and Regionalism in Japan's Foreign and Security Policies', in B. Edstrom (ed.), *Japan's Foreign and Security Policies in Transition* (Stockholm: The Swedish Institute of International Affairs, 1996), pp. 41–60.
2. On the various approaches to 'regions' and 'regionalisms', see A. Hurrell, 'Explaining the Resurgence of Regionalism in World Politics', *Review of International Studies*, 21, 4 (1995), pp. 331–58, and also A. Hurrell, 'Regionalism in Theoretical Perspective' in L. Fawcett and A. Hurrell (eds),

Regionalism in World Politics (New York: Oxford University Press, 1995), pp. 37–73.
3. For a fuller discussion, see R. Higgott and R. Stubbs, 'Competing Conceptions of Economic Regionalism: APEC versus EAEC in the Asia Pacific', *Review of International Political Economy*, 2, 3 (1996), pp. 516–35, and G. Hook, 'Japan and Contested Regionalism', in I. Cook, M. Doel and R. Li (eds), *Fragmented Asia: Regional Integration and National Disintegration in Pacific Asia* (Aldershot: Avebury, 1996), pp. 12–28. For earlier proposals on an 'East Asian' grouping, see P. Korhonen, *Japan and Asia Pacific Integration. Pacific Romances 1968–1996* (London: Routledge, 1998), pp. 176–8.
4. For further details on the emergence of EAEC, see G. Atan, 'East Asian Economic Regionalism. Moving to the Next Level', in K. S. Jomo (ed.), *Japan and Malaysian Development. In the Shadow of the Rising Sun* (London: Routledge, 1994), pp. 326–34.
5. The possible members have included ASEAN, Japan, South Korea, China, Hong Kong, Taiwan. On the Chinese support for the Group, see S. Leong, 'The East Asian Economic Caucus: Regionalism Almost Denied', paper prepared for project on 'new regionalism' by the World Institute for Development Economics Research (WIDER), Helsinki, Finland, 1996, esp. pp. 5–7.
6. There have been clear differences in Japan between more open support for EAEC by the Ministry of International Trade and Industry to occasional outright resistance by the Ministry of Foreign Affairs and political leaders like Prime Minister Miyazawa. See R. Nakayama 'EAEC Kōsō no Keika Bunseki' (Tokyo: MA Dissertation, Graduate School of Law and Politics, University of Tokyo, 1994).
7. *Bōeki to Kanzei* (April, 1996), p. 26.
8. For an explanation and critique of the 'flying geese' model, see P. Korhonen, *Japan and the Pacific Free Trade Area* (London: Routledge, 1994).
9. K. Fujiwara, 'Contending Orders in East and Southeast Asia: Japanese and Asian Perspectives', *Discussion Paper Series* (Tokyo: Institute of Social Science, University of Tokyo, 1994).
10. Also see J. Baker and T. DeFrank, *Politics of Diplomacy: Revolution, War and Peace* (New York: Putnam, 1995), esp. p. 610. The opposition of the US has softened somewhat in the interim. See Leong, op. cit. For broader discussion on APEC and EAEC, see G. Hook, 'Japan and the Construction of Asia Pacific', in A. Gamble and A. Payne (eds), *Regionalism and World Order* (Basingstoke: Macmillan, 1996), pp. 169–206.
11. World Bank, *The East Asian Miracle. Economic Growth and Public Policy* (Oxford: Oxford University Press, 1993), p. xvi.
12. On the use of 'race' in the EAEC debate, see Leong, op. cit, where he cites J. Baker as stating 'I simply could not support an East Asian Caucus that excludes Caucasians', p. 17.
13. For details, see T. Tsubouchi, *Ajia Fukken no Kibō, Mahathir* (Tokyo: Aki Shobo, 1994), p. 149ff.
14. On the 'Japan problem', see K. van Wolferen, *The Enigma of Japanese Power* (New York: Alfred A. Knopf, 1989).
15. Quoted in M. Othman, 'Regionalisation of East Asia: with Particular Reference to Japan and East Asia Economic Grouping', PhD Dissertation, Department of Sociology, University of Sheffield, 1994, p. 251.

16. Quoted in *Bōeki to Kanzei* (July 1996), p. 29.
17. Tsubouchi, op. cit., p.148.
18. On the redefinition ('reconfirmation') of the security treaty, see Hahei Chekku Henshū Iinkai (ed.), *Nichibei Anpo 'Saiteigi' o Yomu* (Tokyo: Shakai Hyōronsha, 1996). On the new guidelines, see *Sekai* Bessatsu (October 1997).
19. On the link between the shrine and militarism, see M. Miyata, 'The Politico-religion of Japan: the Revival of Militarist Mentality', *Bulletin of Peace Proposals*, 13, 1 (1982), pp. 25–30.
20. On the textbook issue, see C. Rose, *Interpreting History in Sino-Japanese Relations* (London: Routledge, 1998), and Y. Irie, 'The History of the Textbook Controversy', and K. Nishio, 'Rewriting Japanese and World History', in special section, 'Behind the Textbook Flap', *Japan Echo* (August, 1997), pp. 34–44.
21. For details, see the Official Report of R. Coomaraswamy, UN Special Rapporteur on Violence Against Women, as submitted to the Human Rights Commission. E/CN/4/1996/53/Add. 1, especially p. 4.
22. *Far Eastern Economic Review* (24 August 1995), p. 37.
23. B. Cumings, 'The Origins and Development of the Northeast Asian Political Economy: Industrial Sectors, Product Cycles, and Political Consequences', in F. Deyo (ed.), *The Political Economy of the New Asian Industrialism* (New York: Cornell University, 1987), especially p. 62.
24. Quoted in J. Nishikawa, 'Japan's Option: International Order or Regional Order', *Peace Research in Japan 1977–78* (Tokyo: Japan Peace Research Group, 1978), p. 19.
25. On the 'Look-East' policy, see K.S. Jomo, 'Introduction' in K.S. Jomo (ed.), *Japan and Malaysian Development. In the Shadow of the Rising Sun* (London: Routledge, 1994), pp. 3–10.
26. For an overview, see K. Onoda, 'Nichibei Bōeki Masatsu to Nihon no Shijō Kaihō', in T. Aoki and K. Umada (eds), *Nichibei Keizai Kankei: Arata na Wakugumi to Nihon no Sentaku* (Tokyo: Keiso Shobo, 1996), pp. 173–204.
27. M. Mahathir and S. Ishihara, *'No' to Ieru Ajia* (Tokyo: Kobunsha, 1994).
28. S. Ishihara, *The Japan That Can Say 'No'* (New York: Simon and Schuster, 1991). Before the publication of this English version, a *samizdat* of the Japanese version of the book, co-authored with Sony's then president, A. Morita, was unofficially translated and distributed to policy-makers and others in Washington by the Defense Advanced Research Projects Agency.
29. See Chapter 8.
30. *Bōeki to Kanzei* (July 1996), p. 32.
31. Fujiwara, op. cit., pp. 21–2.
32. M. Mahathir, *Building a New East Asia*, The Perdana Papers (Kuala Lumpur: The Institute of Strategic and International Studies, 1997). Quoted in *Bōeki to Kanzei* (July 1996), p. 34.
33. N. Gallant and R. Stubbs, 'APEC's Dilemmas: Institution-Building around the Pacific Rim', *Pacific Affairs*, 70, 2 (1997), pp. 203–18.
34. For a different view, see S. Huntington, 'The Clash of Civilizations?', *Foreign Affairs*, 72, 3 (1993), pp. 22–49.
35. For a discussion of these trends as a form of 'new medievalism', see A. Gamble, 'The New Medievalism', paper presented at the Kobe Seminar on Regionalism, Kobe, Japan, 13–14 March 1997.

11 Conclusion

Ian Kearns and Glenn Hook

As stated in our Introduction, the motivation for this volume was the attempt to add something of both theoretical and empirical value to the existing literature on regionalism. Readers will obviously judge for themselves whether we have delivered on this ambitious objective but it is our hope, at least, that by focusing in particular and really for the first time in a single volume upon what we have throughout referred to as the phenomenon of subregional cooperation, *Subregionalism and World Order* has helped to fill something of a gap in the existing treatment of this issue. That it was important to try is, perhaps, most easily verified by the fact that the member states of one subregional grouping or another now cover much of the globe. Consequently, in our judgement, it has never seemed more appropriate, nor perhaps more urgent, for an attempt at understanding subregionalism's emergence and significance to be made. Something is clearly changing and few have so far been addressing what it might be. What, in essence then, does subregionalism represent? How does it relate to the prevailing trend towards a globalization of the world economy and indeed, to the putative, higher level, triadic regionalism often said to be emerging around the US, the EU and Japan? Should subregionalism be welcomed, or should its likely consequences be warned against? Each of the authors writing in the preceding chapters has, of course, attempted answers to these and to other important questions which relate to their own area of regional specialism. Our task here, however, is to draw out some of their individual insights and to step back and sketch out something of the current overview as we see it. The observations which follow are divided into those concerned with a summary of empirical evidence, on the one hand, and those which indicate briefly the theoretical implications which may follow from it, on the other. At the empirical level, it is fair to say that among the studies of the various subregional groupings which have been presented, certain features have emerged as common to all, while others appear as marked and potentially important differences between them. Each of these is dealt with in turn in the section below.

SUBREGIONALISM: A SUMMARY OF EMPIRICAL EVIDENCE

Perhaps one of the most striking features of the subregional projects dealt with in this book is their multidimensional nature, both in terms of their actually existing substance, and in terms of their range of ambition. Subregionalism is far more than simply a series of attempts at economic cooperation. It also, as the evidence presented throughout the book has shown, contains an awareness and an important focus upon security issues, cultural cooperation, and upon the extent to which the politics of identity can be linked to questions of political economy. There are few, if any, examples in existence of an ongoing subregionalist project which, explicitly at least, seeks to limit itself to only one aspect of this agenda. Indeed there is a sense in which subregionalism, as a formal attempt at cooperation by those involved, is an expression of the interrelated nature of all of these issues and of the belief that, where cooperation in each of them has not been present in the past, it needs to be promoted much more systematically in future. Consequently, in sketching out the common characteristics of today's subregionalism one needs to bear two points in mind. First, it is important to do justice to the multidimensional character of the activity involved and second, one must be careful not to confuse the future ambitions of ardent subregionalists with evidence of already existing cooperation. One must be clear, in short, about what the subregionalist projects examined here really are and also about what they are not.

In relation to economic activity, the most obvious point to make is that all of the groupings discussed show a clear commitment to what might be termed 'open subregionalism' meaning that the visibly increased and increasing levels of inter-state cooperation are obviously not motivated by an attempt to construct new power blocs which can then in turn sit behind a firewall of protectionist measures. Trade and cooperation agreements, an important component of the cooperative efforts emerging, are clearly based upon the recognized need to liberalize economic activity and to go for export promoting, rather than import-substituting, strategies of development. This is true, despite the fact that strategic, and not totally 'free' trade, is the more evident ambition. The economic rhetoric of subregionalism, moreover, wherever it is being practised around the world, is clearly rhetoric of a western-led and western-backed variety which emphasizes the need to root out inefficiencies in order to better attract capital and better deal with the demands of growing global competition. What economic subregionalism cannot currently be said to be, however, is an outgrowth of increasing levels of economic regionalization on the ground since, in most cases at least, the subregionalist projects themselves represent the attempt to promote regionalization where it does not yet exist. This would certainly be true of CEFTA and the Black Sea Economic Cooperation Scheme in Europe, ECOWAS in Africa, the ACS in the Caribbean and much of the subregional activity in South America.

In terms of the relationship between security issues and subregionalism, it is striking that security itself is no longer defined simply in terms of military threats from neighbouring states. At least one key feature of the new subregionalism appears to be the attempt to come to terms with a whole new range of transnational security concerns, such as transnational organized crime, drug trafficking, terrorism and the ever-increasing flows of illegal migration. Each of these now represents a new kind of challenge to the authority of state governments and to stability on the territories over which they rule. Each also can be handled far more effectively when the authorities on different sides of a common border strive to cooperate. None of this means, of course, that traditional military concerns have disappeared altogether. ASEAN and the ARF (the security body which has arisen out of it), together with the Malaysian-led EAEC project, for example, are clear evidence that threats of traditional military conflict continue to exist and that some subregionalist projects, at least, have been conceived as part of a broader attempt to ensure that traditional rivalries do not resurface. Still, and even bearing this in mind, it could not be said that subregionalism, even when limited only to security issues, is simply a new solution to a very old security problem. Its character clearly reflects the fact that the perception of new problems also is generating new solutions. This might further be said to extend to the broader approach to foreign policy which is being displayed by many of those states now associated with subregional cooperation. It appears that much more than in the past, smaller or weaker states are attempting real and meaningful coordination of their positions in the international arena in the hope of punching above their weight as individual states across the whole range of the issue agenda.

As with the larger scale regional bloc projects surrounding the EU, Japan and the US, subregional cooperation today clearly also represents a process which is most keenly attractive to policy-makers themselves. It could not be said of any emerging subregional grouping dealt with in this volume that its existence represented the political manifestation of either popular pressure or even widespread public awareness of cooperative efforts. Subregional cooperation is fundamentally an elite-led process wherever one looks around the world and, indeed, as will be discussed later, it is often used in its own right to outmanoeuvre and stifle popular opposition to the kind of politics and neo-liberal political economy which it itself represents. Not surprisingly too, the lack of popular interest and support for subregionalism further reflects an even more fundamental aspect of its nature, namely that despite the clear ambitions to promote cultural affinity and common identity which form an almost standard part of the rhetoric surrounding the creation of subregional projects, there is very little evidence to suggest that new identities are challenging old, or that cultural barriers and stereotypes are being broken down. In East Asia, a regional identity set out in contrast to the cultural traditions of the West is often implied but, as Dominic Kelly's chapter has argued, this does not in fact make it a reality. Indeed, with the possible exception of the Greater China subregion, where, as Ngai-Ling Sum has shown, the impact of previous decentralizations of power may ultimately lead to far stronger regional identities and even to cross-border regional identities between parts of Southern China, Hong Kong and Taiwan, the prevailing feature of subregional cooperation in terms of identity has not been moved to a shared identity, but has been the extent to which most subregional groupings have arisen despite a backdrop of at least one major conflict over the nature of the identity which the group itself should seek to reflect and represent. Even within the Greater China subregion, the possibility of a new regional identity must struggle alongside at least two other possible definitions of what the recent wave of cooperation represents. On the one hand, the central authorities in China see cooperation as an opportunity to reunite China around the shared identity of Chinese nationhood. On the other, the western-influenced Taiwanese, in conjunction with the US and alongside the residue of British colonial rule in Hong Kong, wish to construct subregionalism in the area as a lever to open up the rest of China to western, liberalizing influences. When one looks further afield, one finds only further evidence to suggest that subregional groupings are in fact new venues for the playing out of identity struggles rather than signs that these struggles are themselves being overcome. In Africa, ECOWAS is riven by divisions between Anglophone and Francophone member states and, in the Caribbean, the ACS labours under the weight of the divided Hispanic and Anglophone colonial legacy. In other cases too, such as in the Black Sea Economic Cooperation Scheme and in MERCOSUR, states are attempting to build cooperation not just against the backdrop of unique identities but also in the context of open and quite recent hostility between key member states: Greece and Turkey in the former, for example, and Argentina and Brazil in the latter. While subregionalism represents practical economic cooperation and laudable attempts to better manage the security environment, therefore, one cannot as yet claim that it is a serious means to build new, unified identities through a process of genuine cross-cultural fertilization and exchange: histories and identities remain in tension regardless of the subregionlist project which one chooses to examine.

When we turn to examine aspects of difference among the various subregional groupings discussed in this volume, two points stand out in particular. First, one should perhaps say something about how varied the levels of institutionalization actually are. Some subregional groupings such as CEFTA, for example, have barely become institutionalized at all, while others, like the Black Sea Economic Cooperation Scheme, have on paper at least a fully worked out institutional structure covering meetings between heads of government, ministers, parliamentarians and even a permanent secretariat. These differences in structure are also not limited to the European/Central Asian arena. The ACS has gone for formal institutions, MERCOSUR and the Chilean leadership have not. ASEAN has a formal institutional structure, while subregional cooperation in the Greater China Subregion rests only on informal networks.

Having said this, however, it would be wrong to conclude that the level of institutionalization is a guide to how deep or how effective the resultant cooperation will actually be. If anything, for example, CEFTA has been more effective at promoting cooperation than the Black Sea Economic Cooperation Scheme. Levels of institutionalization are varied, then, but appear not to be of fundamental importance to the cooperation which actually develops. It seems, furthermore, that where ambitious institutional structures have been set out on paper, they often slide into misuse or become an additional source of potential tension between the states involved. Arguments over personnel, or over the location to be chosen as the home for the permanent secretariat are to be found in subregionalism's make-up on every continent.

The second area where one can see a great variety among, and even within, subregional groupings concerns the regime types which appear to be involved. The cases discussed throughout this volume cannot, for example, be said to support the view that democratic regimes are more likely to cooperate with one another than are dictatorships. To be sure, in various subregions a process of democratization certainly has helped to generate a climate in which cooperation could be attempted, Eastern Europe and the territories of the former Soviet Union being obvious examples. There are plenty of other cases, however, most obviously in East Asia and in west Africa, where subregional cooperation is not only underway but also is actually being led by regimes of a less than democratic variety. The existence or otherwise of democracy, therefore, does not appear to be a key variable in the explanation of subregionalism's emergence.

SUBREGIONALISM AND GLOBAL ORDER: THEORETICAL IMPLICATIONS

Having described the core features of subregionalism and having drawn out common features as well as those upon which there are some elements of difference, we now turn to an exploration of the ways in which the subregionalist phenomenon can best be explained. It is not our intention here, however, to engage in a detailed debate of the pros and cons of each of the frameworks set out in the Introduction. Rather, we simply wish to summarize the type of explanation which we believe the evidence presented in this book would tend to support. Consequently, as adumbrated in the Introduction, we should make clear that our explanations of subregionalism draw their inspiration implicitly from a 'new IPE' perspective and from writers such as Robert Cox and Susan Strange in particular.[1] The attraction of this approach is not that it can provide a single explanation of subregionalism whenever and wherever it occurs, but that it sets out a framework which allows the right breadth of questions to be asked. More particularly, explaining subregionalism requires a multilevelled analysis, a capacity to explore historical, economic and ideological structures, and the ability to incorporate both ideas as well as the material world into its explanatory propositions. The new IPE has in many ways been designed explicitly to provide each of these qualities. Consequently, we sketch out below the various levels of analysis needed by dealing with the global, regional and internal state developments which together have provided the essential context for subregionalism's emergence. In doing so, we also cast our net wide to include the relevant economic, ideological and security developments at each of these levels of analysis.

Understanding the nature and direction of change, both globally and within individual states, is the key to an explanation of subregionalism. The world order has gone through a series of transformations in the last decade, some dramatic, some more evolutionary, which together have altered fundamentally the ideological, political and economic contexts within which all actors must operate. At the global level, the end of the Cold War, the collapse of the Soviet Union and the end of US hegemony, themes which have been recurring throughout this book, have all been crucial. The security structures and constraints which, for decades effectively tied most states in the international system down in foreign policy terms, have disappeared in the 1990s. Russian and, to a lesser extent, US retrenchment has opened up space for new patterns of international relations to emerge. Subregionalism is one of these patterns.

Running alongside and, as Susan Strange has argued, perhaps even underpinning these fundamental changes, has been the equally transformative impact of the onset of a globalization of the world economy.[2] Changes in technology and communications have resulted in a massive increase in cross-border, transnational, production processes and have allowed a near total integration of the world's financial markets. Transnational corporations have engaged in a huge expansion of flows of foreign direct investment, now more confident than ever that distance need be no barrier to effective management and integration of corporate functions. There has been a major, consequent, increase in the structural power of internationally mobile capital and this has further given rise to a transnational managerial class which operates, by definition, beyond the control of any single state authority. The whole process, meanwhile, is legitimized by reference to the continuing ascendancy of the neo-liberal ideas originally launched during the Reagan and Thatcher years of the 1980s.

The breadth of such change at the global level has also been reflected in, and reinforced by, certain notable changes in the field of domestic political economy, particularly in the developing world. In Latin America, for example, years of failed attempts at development through import-substituting strategies were ready to give way from the 1980s on to an emerging belief that other, more open, strategies should be tried. The end of the Cold War, furthermore, also ensured that far more serious pressures for democracy were evident in the 1990s, especially within those states in Latin America and East Asia which were ruled by the non-communist dictatorships more easily tolerated at the height of the East–West struggle.

This overall outline of important aspects of change is central to our discussion as it is our contention that such change sits at the root of subregionalism's emergence. In different cases, the strength of influence of each aspect of change will differ, but generally speaking the differences are to be seen as matters of degree and not of fundamental substance. For the lines of influence from the end of the Cold War struggle, from globalization, and from changes in domestic political economy to the advent of subregional cooperation are clear for all to see in each and every case.

Consider, for example, the impact of the end of the Cold War on the dynamics of regional international relations. In Europe, this could hardly have been greater since the old East–West balance of terror has given way to a situation in which the EU is now clearly the dominant force on the continent. In East Asia, too, the combination of an absence of real Russian involvement and a degree of US retrenchment, particularly on security issues, has meant the possible return of a Chinese–Japanese rivalry. Even in the Western hemisphere, where the US has always been dominant, if not hegemonic, it is clear that the Bush and Clinton administrations have used the opening provided by the end of the Cold War to reshape their policies and to use the core issues of trade, aid and democratization to renew and refocus the direction of US influence over Latin America and the Caribbean. In short, the US has been attempting to some extent to replace its global hegemony with a new form of regional domination in its own 'backyard'.

What we now have then, it seems, is a clear shift away from the days of a struggle for global power and a palpable reintroduction of a period when regional hegemons have the freedom to manoeuvre. There is, furthermore, an obvious relationship between the notion of regional hegemony and the occurrence of growing levels of subregional cooperation. On the one hand, subregionalism must be seen at least in part as the response of weaker states within each region to the reality of the new post-Cold War situation. Consider, for example, as Ian Kearns and Gerasimos Konidaris have done, the fact that the whole of the CEFTA membership, and that of the Black Sea Economic Cooperation Scheme, is primarily motivated into subregional cooperation by the simple desire to court the EU. The states involved need the EU badly for trade, aid, and as a source of capital, and fear the total exclusion which otherwise might afflict them if they do not attempt to cooperate in an emerging world of higher level triadic regionalism. Mutual cooperation, so it is thought, can have the advantage of both maximizing the benefits which may be extracted from the EU now and may, in the longer term, prove the suitability of the states involved for eventual full EU membership. Similarly, as Tony Payne and Paul Cammack have argued, initiatives such as the ACS and MERCOSUR must also be seen, at least in part, as obvious attempts to avoid total economic marginalization by, or incorporation on American terms into, the US-led NAFTA project. For EAEC and ASEAN, the picture is slightly more complicated. As both Glenn Hook and Dominic Kelly have shown, one aspect of the motivation for these projects was the hope of generating greater influence on Japan as a possible lever to limit US pressure and involvement in the economic structures of the region. Given the re-emergence of China as a powerhouse of economic growth, however, the aim also has become one of attempting the multilateral management of the potential Japanese–Chinese rivalry. Finally, in the one case of subregionalism from the African continent which has been explored in this book by Stephen Riley, namely ECOWAS, there is again a visible relationship between the notion of regional hegemony and subregional cooperation though, in this case, the subregionalist project itself is a project of the hegemonic power in the area: without Nigerian initiative and support, ECOWAS would probably not continue to exist.

Having noted the relationship between regional hegemony and subregional cooperation, however, we should be careful here to point out that the dominant structure of the contemporary international system is not one of regionally hegemonic powers in competition with one another for global power. To draw such a conclusion would be to concentrate too much on a neo-realist style state-centric level of analysis. States are not best viewed as unitary actors but, to borrow Robert Cox's term, are 'state/society complexes' which reflect a particular distribution of political, economic and ideological forces within a given space.[3] Consequently, the existence of regional hegemons is itself subsumed within the fact that all of the elites involved, either those in hegemonic states or in the smaller and weaker states which surround them, are attempting to react to the pressures and constraints of the more fundamental global financial and production structures and to operate within the limits prescribed by the globally dominant neo-liberal ideology.

This becomes clearer when one considers that the increased structural power of internationally mobile capital, which has emerged as a consequence of globalization, has further meant that state institutions have become more vulnerable to the whims of transnational capital and therefore find it increasingly difficult to pursue any separate national economic development strategies. Again, too, this global level development is vital to an understanding of subregionalism. One clearly could not, for example, explain the obvious commitment of each subregional grouping studied here to an export-led development strategy, open regionalism, and to at least a lip-service commitment to liberal democratic forms of politics without some reference to this global economic context. The structural power of internationally mobile capital, the ideological ascendancy of neo-liberalism and the demands for structural adjustment made by the Bretton Woods institutions amount to a very powerful series of pressures for state conformity with what are now seen as these norms of international behaviour on the economy, democracy and human rights. Perhaps even more fundamentally still, the very speed of the information technology revolution, the sheer scale of investment required to indulge in each of its stages of research and development, and the mobility of the capital required to undertake it has given a premium to the largest units of jurisdiction in the global order. Investors in technology need guaranteed access to markets to ensure the opportunity to recover the vast sums involved in each phase of investment and, for the smaller and/or weaker states trying to survive in this context, one answer is to group together in larger regional or subregional units.

Despite these compelling reasons for emphasizing the global and regional factors which clearly have underpinned the drive to subregionalism one must also, as stated earlier, take on board the often crucial issues relating to the domestic environments of the states involved. For the elites which are pushing subregionalism are not simply doing so because they are forced into this position by external constraints. They are also promoting subregionalism as a method to reinforce their domestic positions given the growing pressures for change which increasingly they are forced to face. In many of the preceding chapters, indeed, powerful arguments have been made to show that subregionalism, in its present form, is popular precisely because it serves the interests of certain state and economic elites. Consider, for example, the cases of Brazil and Argentina, the leading states in MERCOSUR where, as Paul Cammack has pointed out, subregional cooperation and the externally imposed economic disciplines which it involves have provided crucial legitimation for governments which, although isolated, are pushing for a liberalization of the economy and a retrenchment of the state. Similarly, in her analysis of Chile, Jean Grugel pointed out that the perceived necessity of the Chilean technocratic and business elites' response to its external environment has usefully served to de-politicize many aspects of economic liberalization within the state.

Other examples, this time from Africa and East Asia, though in ways influenced specifically by their own cultural and economic histories, can further serve to reinforce this point. In ECOWAS, for example, there is little doubt that subregional cooperation has been pushed by the Nigerian leadership precisely because it served the domestic interests of the Nigerian regime. In this case the objective was to ensure that elements of the military which may be interested in coup-plotting were kept busy outside of the country performing the peace-keeping duties which the ECOWAS arrangements required. If those same elements of the military could be materially satisfied by the profits of looting from the peoples they were supposed to be protecting then all well and good. For EAEC and ASEAN (once the claim that they are based upon a unique, unified and anti-liberal set of Asian cultural values is dismissed) subregionalism can also be revealed at least in part as an attempt to mask the poor human rights and undemocratic track records of several of the regimes in the area. While there are powerful reasons at the global level as to why subregionalism should exist in East Asia, therefore, an explanation of it would clearly be incomplete without this reference to the fact that it further protects domestic regimes against rising internal pressures for greater openness in politics and greater fairness in the economy. While in general, and as already shown earlier, the external pressures and constraints are real enough then, the point being made here is that domestic elites in favour of accommodating external conditions are actually exaggerating the extent to which there is no room for manoeuvre in order to deliberately limit potential domestic opposition from those who may stand to lose. Subregional cooperation is not a neutral in political and economic terms. It is very definitely being constructed in such a way as to allow elites to make their way in the changing game of global capitalism.

This, of course, has very profound implications for our ultimate assessment of contemporary subregionalism. Here, we can perhaps do no better than to reassert the somewhat pessimistic comments of Andrew Gamble and Anthony Payne in the concluding chapter of *Regionalism and World Order*, the book which precedes this one as the major output of the research cluster on regionalism at the Political Economy Research Centre, The University of Sheffield.[4] The present stage of the world order is characterized by its complexity and it is by no means clear what the process of ongoing change will mean for the existing sites of its governance. What is clear, however, is that change is creating its own patterns of winners and losers. Subregional groupings of weaker states and institutions, if managed in an enlightened way and if opened up to the wider influences and interests of labour and of civil society more broadly, may yet play a constructive role in helping to mitigate the downside of globalization. If not, and if subregionalism continues to use the public face of international cooperation to mask the needs of a few private interests, particularly amongst the elite, then it may well become one of the targets for any future radical challenge to capitalist civilization. There is a way to go yet before today's subregionalism can be wholeheartedly welcomed.

NOTES

1. The major works to be referred to here would be S. Strange, *States and Markets* (London: Pinter, 1988); and R. Cox, *Production Power and World Order: Social Forces in the Making of History* (New York: Columbia University Press, 1987).

2. For an account of her argument in relation to globalization and the collapse of the Soviet Union, see S. Strange, 'States, Firms and Diplomacy', *International Affairs*, 68, 1, 1992.
3. See R. Cox, 'Social Forces, States and World Orders: Beyond International Relations Theory', *Millennium, Journal of International Studies*, 10, 1981.
4. See A. Gamble and A. Payne, *Regionalism and World Order* (Basingstoke: Macmillan, 1996), pp. 247–64.

Index

Abacha, Sani, 77, 80
Abiola, Moshood, 77
Abuja, Treaty of, 70
ACS, 10–11, 92, 117–18, 251
 aims/evaluation, 133–6, 249, 254
 background/origins, 120–30, 250
 Convention, 130–33
African Economic Community (AEC), 9, 65, 70, 82
AFTA, 166, 172, 181–4, 191, 194n29
Albania, 53–5
 and BSEC/BSTD, 42, 49–50, 55, 61n24
Alfonsín, Raúl, 95, 102, 106
All People's Congress (APC), 74
Alpe-Adria community, 22–4
AMCHAM, 153
Americas Summits, 99, 142
Andean Pact, 93, 141–2
Anglo-Malayan Defence Agreement, 172, 185
Anguilla, *118*, 130
Ani, Anthony, 68
Aninat, Eduardo, 148, 152–3
Antigua and Barbuda, 117–18
APC (All People's Congress), 74
APEC, 5, 139, 165–6, 174, 183
 and Chile, 139, 155
 and EAEC/EAEG, 167, 223–4, 227, 242
 and US, 174, 189, 231
Aramburu, Pedro Eugenio, 108
ARATS, 209
ARENA, 106
ARF, 165, 172, 179, 186–91, 193n12, 249
 and 'ASEAN way', 175, 190–1
 and Japan/China, 187–91
 and US, 174, 187–9, 225
Argentina, 97–8, 100–2
 liberalism, 103, 108–10, 113; failure, 100, 107–8
 and MERCOSUR, 95, 98, 103–5, 108–10, 113, 256

 military rule, 101–2, 105–6, 108–10
 Peronism, 101, 105–6
Aristide, Jean-Bertrand, 126
Armenia, 53–4
 and BSEC, 42, 49–50, 61n24
Aruba, *118*, 130
ASA, 7, 172, 185
Asael, Hector, 155
ASEAN, 11, 169–75, 182, 241, 256–7
 and EAEG/EAEC, 223–4, 231, 238–9
 Free Trade Area, *see* AFTA
 and ideology, 7, 175
 institutionalization, 173–4, 251
 and Japan, 176–9, 181, 238–9, 254–5; *see also* ASEAN plus-three
 Regional Forum, *see* ARF
 and security, 165, 170–71, 178–9, 185–6, 238–9, 249
 and US, 174–6, 178, 181
 and Vietnam, 169, 171, 178–9, 238
 see also 'ASEAN way'
ASEAN plus-three, 224–5, 231, 239–40, 242
'ASEAN way', 11, 173, 191
 and economics, 179–84, 191–2
 origins, 175–9
 and politics, 174, 184–91
 social costs, 12, 184, 192
Asia–Europe Meetings (ASEM), 13, 174, 225
Asian Collective Security System, 184–5
'Asia-Pacific': identity, 223, 242
 and US, 227, 231
Asia Pacific Economic Cooperation, *see* APEC
Association for Relations Across the Taiwan Straits, 209
Association of Caribbean States, *see* ACS
Association of Southeast Asia (ASA), 7, 172, 185
Association of Southeast Asian Nations, *see* ASEAN
Asunción, Treaty of, 95
Australia, 172, 230
Austral Plans (Argentina), 102

Index

Austria, 23, 43
Aylwin, Patricio, 146, 148, 151
Azerbaijan, *53–4*
  and BSEC, 42, 49–50, 61n24

Babangida, Ibrahim, 77, 81
Bahamas, 117–18
Baker, James, 227
Bakos, Gabor, 29, 35
bananas: 1992 dispute, 128
Barbados, 117–18
Basic Labour Law (Taiwan), 202
Bayart, J.-F., 71
Belize, 117–19
Benin, *64*, 68
Berger, M., 178
Bermuda, *118*, 130
Black Sea
  Strategic Action Plan, 46
  study centre: plans, 47
  subregionalism, *see* BSEC; BSTD
Bolivia: and MERCOSUR, 113
Bosnia-Hercegovina, 23, 43
'Bosphorus Statement' (BSEC), 42
BRASS, 45
Brazil
  Democratic Movement (MDB), 107
  economy, 96, 101–4
  Labour Party (PTB), 105
  and MERCOSUR, 95, 98, 103–5, 108–10, 113, 256
  politics, 105–7, 109; liberalism/democracy, 97–8, 100, 104–10, 113; military rule, 101–2, 106, 109–10
British Virgin Islands, *118*, 130
Brizola, Leonel, 107–8
Brodsky, Joseph, 27–8
Brown, Ron, 152–3
Brunei: and ASEAN, 169
BSEC, 8–9, 18, 249
  areas of cooperation, 43–7, 51
  Bank (BSTD), 49–51, 53–4
  Business Council, 43, 49
  Coordination Centre for Exchange of Statistical Data, 47
  creation/aims, 41–3, 58–9
  and ex-communist member states, 18, 52–6, 58
  external relations, 50–51; with EU, 9, 47, 51, 58–9, 254
  institutions, 9, 47–50, 251
  internal conflicts, 18, 52, 56–8, 61n30, 250
  MMFAs, 44, 47–8
  observers, 43, 50, 60n18
  PERMIS, 48, 61n24
  problems, 51–9
  Subsidiary Bodies (SBs), 48, 51
  Summit Declaration, 42–4, 53, 57–8
BSTD Bank, 49–51, 53–4
Bucharest Convention, 46
Buchi, Herman, 149–50
Bulgaria, 26, 53–4
  and BSEC, 42, 49, 61n24
Bumiputra policy, 234, 237
Burkina Faso, *64*, 72
Bush, George, 91, 98, 123, 127, 143, 254

Cabral, Luiz, 72
CACM, 93, 121, 132, 135, 141–2
  and ACS, 128, 132
'cadre entrepreneurs', 203, 210
Callejas, Rafael, 127
Cambodia, 169, 171, 178–9, 238–9
Cammack, Paul, 10, 95–115, 254, 256
Cape Verde islands, *64*, 67
Cardoso, Fernando Henrique, 98, 107, 112–13
Caribbean Basin, 117, 119–22, 128, 134–5, 136n2
  Free Trade Agreements Act, 129
  Initiative (CBI), 121–2, 129–30
  Technical Advisory Group (CBTAG), 127–9
  see also ACS; CARICOM
Caribbean Commission: proposal, 123
CARICOM (Caribbean Community), 93, 121–3, 125–6, 134–5
  and ACS, 124–5, 128, 131–2
  Bureau, 123–5
CARIFTA (Caribbean Free Trade Area), 121
Castello Branco, Humberto, 108
Castro, Fidel, 136
Cavallo, Domingo, 103
Cayman Islands, *118*, 130

CEAO, 69
CEFTA, 8, 21, 33–9, 249, 251
  Agreement, 24–6, 37
  and EU/West, 26, 31–4, 37, 39, 254
CEI (Central European Initiative),
  23–4, 37, 45, 51
Central American Common Market,
  see CACM
Central American states, 127–9
central Europe, 22, 27–9
  subregionalism, 17, 21–39
  see also CEFTA; CEI
CFA zone, 66, 69, 74–5
Chen Yun, 210
'Chicago Boys': in Chile, 147
Chile
  agriculture, 150–52, 156–7
  and Andean Pact, 142
  and APEC, 139, 155
  and EU, 145, 155
  exports, 145
  liberalism, 11, 99–100, 144–6,
    157–9, 256
  and MERCOSUR, 113, 139, 141,
    143, 154; negotiations, 152,
    155–7
  and NAFTA, 113, 139, 141, 143,
    152–5
  subregionalism, 11, 93, 157; for
    growth, 140, 144–6;
    state/exporters coalition, 146–52,
    155, 157–8, 256
  and US, 145, 152–5
China
  and ARF, 189–91
  and ASEAN, 171
  and EAEC, 224, 237–40, 254–5
  Greater, *see* Greater China
  and Hong Kong, 202, 205–7, 212–14,
    217
  identity struggles, 210–12
  local development–'Greater
    Shanghai', 215–17; southern
    China, 198–200, 202–8, 210–11
  'Open Door' policy, 12, 198–200
  and Taiwan, *see* Taiwan
  and US, 207–9
China International Trust and
  Investment Company, 203

CIEPLAN, 146, 148
C/LAA, 127
Clinton, Bill
  and Hong Kong, 207–8
  and LAC, 10, 129, 143, 154, 254
CMEA, 24, 32, 34–5
Cold War
  and bilateralism, 6–7, 223
  end, 5, 253–4
  and East Asia, 227, 238–41, 254
Collor de Melo, Fernando, 95, 98,
  103–5, 108
Colombia, 117–20, 125
Colon, Rafael Hernandez, 127
colonialism/imperialism
  and Caribbean, 120
  in East Asia, 212–14, 232–4
  and ECOWAS, 66, 71–2
Colorado Party (Paraguay), 111
Common Market of the South, *see*
  MERCOSUR
Communauté Financière Africaine
  (CFA): zone, 66, 69, 74–5
communism: and (east-)central Europe,
  27–8, 32
Concertacion Democratica, 146, 148,
  159
  and business sector, 149–51, 155
*Confederacion de Produccion y
  Comercio (CPC)*, 150, 153
'Copenhagen conditions', 38
Costa Rica, 117–19
Council of Baltic Sea States, 51
Council of Europe, 51
Council of Mutual Economic
  Assistance, *see* CMEA
Cox, Robert, 5, 252, 255
Croatia, 23, 43
Cross, Malcolm, 120
Cruzado Plan, 102
Cuba, 117–*18*, 121–2, 126
Cyprus: and BSEC, 43
Czechoslovakia/Czech Republic, 23,
  31, 38
  and CEFTA, 24–5, 35–6

Daddieh, C., 73
Danube Alliance, 29
Danube Confederation, 29

da Silva, Luis Inacio ('Lula'), 104, 107–8
De Mello, J., 71–2
Democratic Independent Union (UDI: Chile), 150, 157
Democratic Party (Hong Kong), 212
Democratic Progressive Party (DPP: Taiwan), 211, 214–15
democratization, 251–2
Deng Xiaoping, 198, 200, 210
de Riz, L., 100
'Dirty War' (Argentina), 110
Doe, Samuel, 80
Dominica, 117–*18*
Dominican Republic, 117–*18*, 122–3, 126
DPP, 211, 214–15
drug trade
  in Caribbean Basin, 135
  in Nigeria, 75, 79
Duarte, Simón Molina, 133

EAEC, 165, 172, 193n23, 243, 256–7
  as ASEAN plus-three, 224–5, 231, 239–40, 242
  and China, 224, 237–40, 254–5
  as EAEG, 223–4, 237
  and Japan, 223–4, 240, 254–5; Malaysian strategy, 166–7, 226, 228, 231–8, 241–2
  and Malaysia: proposal: reasons, 12–13, 166–7, 223–6, 231–8, 241–2; US/West response, 227–30
  and security, 238–41, 249
  and US/West, 96, 167, 226–31, 242
EAEG, 223–4, 237
EAI, 91, 98, 123, 127, 130
East Asia
  and Cold War end, 227, 238–41, 254
  and colonialism, 212–14, 232–4
  and globalization, 171, 180–82, 186, 210, 230
  identity, 223–4, *see also* 'ASEAN way'
  'miracle' economy, 192, 227, 237
  subregionalism, 165–7, *see also* ASEAN; China; EAEC

East Asia Coprosperity Sphere, 166, 177, 232–3
East Asian Economic Caucus, *see* EAEC
East Asian Economic Group, *see* EAEG
east-central Europe, 22, 27–8
  and EU/West, 32–4, 39
  and NATO, 26, 36–8
  subregionalism: causes, 33–7
ECLAC, 93, 99, 132
ECOMOG, 19, 67, 77, 81
Economic Cooperation Organization, 51
ECOWAS, 9–10, 18–19, 65, 82–4, 249
  aims, 70, 82
  Committee of Five, 67
  creation, 63, 66–8
  French influence, 66, 68–9, 75–6, 82
  Fund for Cooperation, 68
  institutions, 67–70
  and Lagos Treaty, 66–70, 78, 80
  Monitoring Group, *see* ECOMOG
  and Nigeria, *see under* Nigeria
  problems, 65–6, 71, 82–4
  and security, 9, 65, 80–81, 83
  subregion, *see* West Africa
  and trade, 71–6
  and West, 70–71, 81–2
Egypt: and BSEC, 43
Elekdag, Suekrue, 41
elites: and subregions, 249–50, 256–7
El Salvador, 117–19, 130
EMU: and Greece, 55
Enterprise for the Americas Initiative, *see* EAI
Essen Summit, 38
Estonia: and CEFTA, 26
Euro-Mediterranean Initiative, 51
Europe Agreements, 34, 37
European Commission, 45
European Community (EC), 123, 128–9, 225
  as (sub)region, 2, 4, 7
European Union (EU), 7
  and Alpe-Adria community, 22
  and BSEC, 9, 47, 51, 58–9, 254
  and CEFTA/east-central Europe, 26, 33–4, 37–9, 254

Index

and Chile, 145, 155
and EAEC proposal, 225–6
as ECOWAS model, 70
and MERCOSUR, 10, 114
and Turkey, 18
'Executive Outcomes', 79
export orientation, 248
in LAC, 91–2, 100–1, 121
Eyadema, Gnassingbe, 66

Falklands (Malvinas), 102, 109–10
FEDEFRUTA, 150–51, 156
Finland: and EU, 7
Five-Power Defence Arrangements, 172, 185
Foccart, Jacques, 75
Founding Act on Mutual Relations (Russia/NATO), 38
Foxley, Alejandro, 148
France: influence
   in Caribbean, 130, 136
   in West Africa, 66, 68–9, 75–6, 82
Frei, Eduardo, 146, 148, 151
French Guiana, *see* Guyane
Frondizi, Arturo, 106
FTAA (Free Trade Area of Americas), 10, 113, 134–5, 143
Fujian, 198–200, 203
FYROM (Macedonia), 23, 26, 43

G3 (Group of Three), 92, 121, 125
Gambia, *64*, 72–4
Gamble, Andrew, 4, 257
Gamcikovo-Nagymaros: dam, 24, 31–2
Garcia, Alvaro, 156
Garton Ash, Timothy, 27–9
Geisel, Ernesto, 102, 108
General Agreement on Tariffs and Trade (GATT), 26, 128, 225–6
Generalized System of Preferences (GSP), 229–31
Georgia, 42, 49–50, *53–4*, 61n24
Ghana, *64*, 72–4, 81
Gill, Henry, 132
globalization, 2, 4, 247, 253–6
   and East Asia, 171, 180–82, 186, 210, 230
Gorbachev, Mikhail, 36
Gortari, Salinas de, 125

Gowon, Yubuku, 66–7
Gramsci, Antonio, 5
Greater China, 12, 197–8, 251
   development, 198–207, 215–18
   external relations, 207–9
   identity struggles, 166, 209–15, 250
Greater East Asia Coprosperity Sphere, 166, 177, 232–3
Greece
   and BSEC, 42, 49–50, 55, 61n24/30
   and Turkey, 18, 56
Grenada, 117–*18*, 121–2
Group of Three (G3), 92, 121, 125
Grugel, Jean, 11, 91–3, 98–9, 139–61, 256
GSP, 229–31
Guadeloupe, *118*, 130
Guangdong: development, 198–200, 202–3, 208, 211, 216
Guatemala, 117–19
Guidelines for Defence Cooperation (US–Japan), 231
Guidelines for National Unification (Taiwan), 208
Guinea, *64*, 72
Guinea-Bissau, *64*, 72
Guyana, 117–19
Guyane (French Guiana), 118–19, 130

Haiti, 117–*18*, 122–3, 126
Hand of Friendship Agreements, 36
Havel, Vaclav, 31, 37
Honduras, 117–19
Hong Kong: and China
   and democracy, 207–8, 211–13
   entrepôt/finance role, 206–7, 217
   and 'Greater Shanghai', 216–217
   identity struggles, 200–1, 212–14
   southern, 199, 202–6, 208, 217
Hong Kong Science Park, 217
Hook, Glenn, 1–13, 165–7, 223–45, 247–58
Hosokawa, Morihiro, 233
Houphouet-Boigny, Felix, 66, 75
Hungary, 23, 28, 30, 38
   and CEFTA, 24–5, 35–6
   and Gamcikovo-Nagymaros dam, 24, 31–2
Hun Sen, 169

Hurrell, Andrew, 2, 4, 65
Hyde-Price, A., 28

IBSC, 43, 50
Illia, Arturo, 106
import-substituting industrialization, see ISI
Indonesia, 173, 182, 185, 234
  and ASEAN, 169, 171
  institutionalization, 251
Insulza, Jose Miguel, 148, 152, 154
International Black Sea Club, 43, 50
International Trade Commission, 153
Irigoyen, Hipólito, 105
Ishihara, Shintaro, 236
ISI: in LAC, 91, 101, 121, 141, 147, 253
Israel: and BSEC, 43
Italy, 23, 43
Ivory Coast, *64*, 68, 72, 76, 81

Jamaica, 117–*18*
Japan
  and ARF, 187–9
  and ASEAN, 176–7, 179, 181, 254–5; security role, 178, 238–9
  and EAEC, 223–4, 240, 254–5; Malaysian strategy, 166–7, 226, 228, 231–8, 241–2; US opposition, 227, 231
  and US, 176–7, 207, 227–8; security, 179, 229, 231
Japan Sea Zone, 6
Jara, Alejandro, 148
Jawara, Dauda, 66
Jiang Zemin, 216
Jordan: and BSEC, 43

Kamajor militia (Sierra Leone), 79
Kantor, Micky, 129
Kaplan, Robert, 78–9, 82
Kazakhstan: and BSEC, 43
Kearns, Ian, 1–13, 17–19, 21–40, 247–58
Kelly, Dominic, 11–12, 169–95, 250, 254
Klaus, Vaclav, 32
KMT government (Taiwan), 208–9, 214–15

Konidaris, Gerasimos, 8–9, 41–61, 254
Konrad, Gyorgy, 27–8
Korea, South: and EAEC, 224, 231
Korean War, 176
Krakow Summit: and CEFTA, 24, 37
Kubitschek, Juscelino, 101
Kundera, Milan, 27–8

Lacalle, Luis Alberto, 95, 112
LAC (Latin America and Caribbean) subregionalism, 91–3, 141–4; *see also* ACS; MERCOSUR; *and named* countries
  and US, 91–2, 121–2, 140–3; *see also* FTAA; NAFTA
Lagos, Treaty of, *see under* ECOWAS
Lamounier, B., 100
Laos: and ASEAN, 169, 171
Larrain, Felipe, 153
latifundistas: in Chile, 147, 150
Latin American Economic System, 132
Latin American Free Trade Association, 93
League of Arab States, 51
League of Nations: and Japan, 228
Lee Teng-hui, 207–8, 215, 217–18
Leys, C., 71
Liberal Democratic Party
  in Hong Kong, 211
  in Japan (LDP), 232–3
Liberia, *64*, 72, 77, 79–81
Li Peng, 210, 223, 231, 237
Lithuania: and CEFTA, 26
'Little Entente', 29–30
Lomé Convention, 126, 128
Lonardi, Eduardo, 108
'Look East' policy, 235
'Lula' (da Silva), 104, 107–8

Macau: and China, 199–200
Macedonia (FYROM), 23, 26, 43
Madrid Summit (1997), 38
Mahathir Mohamad: and EAEC, 12, 231, 239–40
  and Japan, 166, 226, 231, 233–6
  and West, 227–8, 230
Malaysia, 172, 182
  and ASEAN, 169, 171
  and EAEC, *see under* EAEC

Mali, *64*, 79
Malvinas: invasion, 102, 109–10
Manley, Michael, 125, 127–8
Maoism, 200
Maphilindo, 172–3, 185
Martinique, *118*, 130
Masaryk, Tomas, 29
Mauritania, *64*, 72
MDB, 107
Mekong Basin: development, 224–5
Menem, Carlos, 102–3, 106, 108, 110, 113
 and MERCOSUR, 95, 98, 104–5
MERCOSUR, 10, 92, 95–7, 113–14, 250–51
 and Chile, 113, 139, 141, 143, 154; negotiations, 152, 155–7
 and EU, 10, 114
 and NAFTA, 97, 113, 139–40, 143, 254
 and Paraguay/Uruguay, 95–6, 110–12
 regional context, 98–100; Argentina/Brazil, 95–8, 100–10, 113, 256
Mexico, *118*–19, 120
 and ACS, 117, 135
 and CARICOM, 125–6
 and NAFTA, 91, 98–9, 135, 140, 142
 and Malaysia, 225
 and US, 91–2, 98, 140, 142–3
microregionalism, 6
Mitteleuropa: concept, 27, 29
MOFERT, 203
Moldova, *53*–4
 and BSEC, 42, 49–50, 61n24
Montserrat, *118*, 130
Munoz, Heraldo, 148
Murayama, Tomiichi, 233
Myanmar: and ASEAN, 169, 171

NAFTA, 91–2, 98, 123, 142
 and ACS, 127, 130, 135, 254
 and CBI, 129–30
 and Chile, 113, 139, 141, 143, 152–5
 and EAEC proposal, 225–6
 and EAI, 98, 123, 127
 and FTAA, 113, 134–5, 143

 and G3, 125
 and MERCOSUR, 97, 113, 139–40, 143, 254
 and Mexico, 91, 98–9, 135, 140, 142 and Malaysia, 225
 and US dominance, 97, 140, 143, 154
Nakasone, Yasuhiro, 232
National Democratic Union, 108
National Endowment for Democracy, 100
National Party (Chile), 149
National Renovation Alliance (Brazil), 106
National Renovation Party (RN: Chile), 150, 157
NATO
 and east-central Europe, 26, 36–8
 and Russia, 38–9
Nee, V., 203
Netherlands Antilles, *118*, 130
Neves, Tancredo, 107
New Economic Policy (Malaysia), 237
'new IPE', 4–6, 252
'new' regionalism, 91, 93, 193n12
New Zealand, 155, 172, 230
Nicaragua, 117–19, 121–2
Niger, *64*, 72, 82
Nigeria, 64, 72, 76–8
 and ECOWAS, 9–10, 68–9; exploitation, 19, 64, 78, 82–4, 256; in Liberia, 77, 80–81
 trade, 72–4; in drugs (illicit), 75, 79; in oil, 64, 76–7, 82
Niigata: and Japan Sea Zone, 6
Nike, 220n11
Nixon, Richard, 178–9
Nkrumah, Kwame, 67
North American Free Trade Agreement, *see* NAFTA
North Atlantic Treaty Organization, *see* NATO

OAU, 70
Obasanjo, Olesugun, 77
Odessa Declaration, 46
OECS, 121, 132
Official Development Assistance (ODA: Japan), 233–4
Ominami, Carlos, 148

'Open Door' policy: in PRC, 12, 198–200
'open subregionalism', 248–9
Organization for Security and
  Cooperation in Europe, 51
Organization of Eastern Caribbean
  States (OECS), 121, 132
Oviedo, Lino, 112
Ozal, Turgut, 41

Pacific, *see* 'Asia-Pacific'
'Pacific way', *see* 'ASEAN way'
Panagariya, A., 71–2
Panama, 117–19
Paraguay: and MERCOSUR, 95–6, 110–12
'particularism': and EAEC, 230
Partnership for Peace Initiative, 38
Patten, Chris, 207–8, 212–13
Payne, Anthony, 4, 98–9, 257
  on ACS, 10–11, 117–37, 254
Pearce, J., 158
Pearl River Delta, 199, 202, 217
Pentagonale Initiative, 23
People's Republic of China (PRC), *see* China
Perón, Isabel, 101
Perón, Juan, 101, 105–6, 108
Philippines, 169, 172, 182
Pinochet, Augusto, 11, 144, 147–9
Plaza Accord (1985), 235
PMDB, 107
Poland, 23, 28, 38, 43
  and CEFTA, 24–5, 32, 35–6
Pompidou, Georges, 69
populism: in Argentina/Brazil, 101, 105–6
Prague Summit: and CEFTA, 24–6, 37
PRC, *see* China
Prem Tinsulanonda, 169
PSD, 105–7
PT, 104, 107
PTB, 105
Puerto Rico, *118*, 127–8, 130

Quadros, Janio, 108
quanxi: and Hong Kong/Taiwan, 204–5, 217
'quiet' diplomacy, 171, 173–4, *see also* 'ASEAN way

Radical Party (Argentina), 106
Raina, Roberto, 126
Ramos, Fidel, 230, 239
Ramphal, Sir Shridath, 123
Reagan, Ronald, 121
regionalism
  approaches, 2–4, 223
  levels, 6–8
  'new', 91, 93, 193n12
  and new IPE, 4–6
  and regionalization, 4–5
regionalization, 4–5, 249
  and EAEC, 230
Riley, Stephen, 9–10, 63–87, 255
Rio Group, 114
RN, 150, 157
Rodriguez, Andres, 95, 111
Romania, 26, *53*, *54*
  and BSEC, 42, 49–50, 61n24
Russia/Russian Federation, *53–4*
  and BSEC, 42, 49–50, 56–7, 61n24
  and East Asia, 239
  and NATO, 38–9
  as Soviet Union, 18, 32, 42, 56; and US, 171, 180
  and Turkey, 18, 56–8

St Kitts and Nevis, 117–*18*
St Lucia, 117–*18*
St Vincent, 117–*18*
Salinas, Carlos, 99
Sanguinetti, Julio, 112
Sankara, Thomas, 72
Sarney, José, 95, 102–3, 107
Saro-Wiwa, Ken, 77
Scully, T.R., 100
SEANFWZ, 187, 191
SEATO, 7, 166, 172, 184–5
security, 249
  and BSEC, 18, 52, 56–8, 250
  in central Europe, 36–8
  in East Asia, 172; and ASEAN/ARF, 165, 170–71, 178–9, 184–91, 238–9, 249; and EAEC, 238–41, 249; US/China, 207, 209
  in LAC, 109–10, 135–6
  in West Africa, 9, 65, 78–81, 83

SEF, 209
SELA: and ACS, 132
Senegal, *64*, 72, 76, 79
Seneildín, Mohammed Ali, 110
Senghor, Leopold, 66
SFF, 150, 155–6
Shanghai-Pudong ('Greater Shanghai'), 215–17
Shantou (SEZ), 199
Shenzhen (SEZ), 199
Shevardnadze, Eduard, 57
Shumaker, D., 31
Sierra Leone, *64*, 72–5, 79
Silva, E., 147–8
Singapore, 169, 172, 177–8
Slovakia, 23, 31, 43
  and Hungary/Hungarians, 24, 30–32
Slovenia, 23, 26, 43
SNA (*Sociedad Nacional de Agricultura*), 150–51, 156–7
Social Democratic Party (Brazil), 105–7
Sociedad de Formento Fabril (SFF), 150, 155–6
Sociedad Nacional de Mineria, 150
SONAMI, 150
Southeast Asia Nuclear Weapons Free Zone (SEANWFZ), 187, 191
Southeast Asia Treaty Organization, *see* SEATO
southern China, *see under* China
South Korea: and EAEC, 224, 231
Soviet Union, *see under* Russia
Statements on Dialogue (NATO), 36
Stevens, Siaka, 74
Straits Exchange Foundation, 209
Strange, Susan, 5, 252–3
Stroessner, Alfredo, 111
sub-subregionalism, 6
Suharto, Thojib: and EAEC, 224
Sukarno, Achmad: regime, 185
Sum, Ngai-Ling, 12, 197–221, 250
'Summer' Plan (Brazil, 1987), 102
Summits of the Americas, 99, 142
Suriname, 117–19, 125–6

Taiwan: and China, 200–1, 208–9, 211, 214–15, 217–18, 250
  and EAEC, 224
  trade/investment, 202; in Shanghai area, 216–17; in southern China, 199, 204–7
  and US, 207–8
Tanaka, Kakuei, 234
Taylor, Charles, 80–81
Taylor, L., 149
'technopols': in Chile, 148
telecommunications: and BSEC, 45, 51
Thailand, 169, 172, 178, 182, 234
Togo, *64*
Toure, Sekou, 72
tourism/transport
  and ACS, 133–4, 136
  and BSEC, 44–7
Trinidad and Tobago, *117–18*
  as ACS location, 132–3
Troncoso, Carlos Morales, 126
Tuareg militia, 79
Tubman, William, 67
Tunisia: and BSEC, 43
Turkey, 55
  and BSEC, 18, 42, 49–51, 55, 61n24/30
  and EU, 18
  and Greece, 18, 56
  and Russia, 18, 56–8
Turks and Caicos Islands, *118*, 130

UDI, 150, 157
UDN, 108
UEMOA, 69
Ukraine, *53–4*
  and BSEC, 42, 49–50, 61n24
United Malays National Organization (UMNO), 226
United Nations
  Economic Commissions for Africa, 67; for Europe, 51; for Latin America, *see* ECLAC
  and Taiwan, 217–18
Uriburu, José Félix, 105
Uruguay, 95–6, 110–12

US (United States)
   and APEC, 174, 189, 231
   and ARF, 174, 187–9, 225
   and ASEAN, 174–6, 178, 181
   and China/Hong Kong, 207–9
   and EAEC, 96, 226–31, 242
   and Japan, 176–7, 207, 227–8;
     security, 179, 229, 231
   and LAC, 91–2, 121–2, 140–43, *see also* FTAA; NAFTA; Chile, 145, 152–5; Mexico, 91–2, 98, 140, 142–3
   protectionism, 92, 142–3, 154
   SEATO, 7, 166, 172, 184–5
   and Soviet Union, 171, 180
   and Taiwan, 207–8
   Trade Representatives Committee (USTR), 153
US–Chile Chamber of Commerce, 153
US Virgin Islands, *118*, 130

Valdes, Juan Gabriel, 148, 152, 154, 156
van Klaveren, Alberto, 148
Vargas, Getulio, 101, 105, 108
Venezuela, 117–20
   and CARICOM, 122–3, 125

Videla, Jorge Rafáel, 102, 108
Vietnam: and ASEAN, 169, 171
   and Cambodia, 178–9, 238
Virgin Islands, *118*, 130
Visegrad Summit: and CEFTA, 24, 37

Walesa, Lech: on CEFTA, 31
Warsaw Pact: and CEFTA, 24
Wasmosy, Juan Carlos, 95, 111–12
West Africa
   cultures, 70–71
   economies, 63–4, 70
   politics, 72–3
   security, 9, 65, 78–81, 83
   trade, 72–5
West Indian Commission, 123–4
Workers' Party (Brazil), 104, 107
World Trade Organization, 44

Xiamen (SEZ), 199

Yopo, Boris, 148
Yugoslavia, 23, 43

Zartman, William, 82
Zhuhai (SEZ), 199